T0361001

PROJECT MANAGEMENT FOR RESEARCH

A Guide for Graduate Students

Industrial Innovation Series

Series Editor
Adedeji B. Badiru
Air Force Institute of Technology (AFIT) – Dayton, Ohio

PUBLISHED TITLES

Carbon Footprint Analysis: Concepts, Methods, Implementation, and Case Studies,
Matthew John Franchetti & Defne Apul

Cellular Manufacturing: Mitigating Risk and Uncertainty, *John X. Wang*

Communication for Continuous Improvement Projects, *Tina Agustiady*

Computational Economic Analysis for Engineering and Industry, *Adedeji B. Badiru &
Olufemi A. Omitaomu*

Conveyors: Applications, Selection, and Integration, *Patrick M. McGuire*

Culture and Trust in Technology-Driven Organizations, *Frances Alston*

Design for Profitability: Guidelines to Cost Effectively Management the Development Process
of Complex Products, *Salah Ahmed Mohamed Elmoselhy*

Global Engineering: Design, Decision Making, and Communication, *Carlos Acosta, V. Jorge Leon,
Charles Conrad, & Cesar O. Malave*

Global Manufacturing Technology Transfer: Africa–USA Strategies, Adaptations, and Management,
Adedeji B. Badiru

Guide to Environment Safety and Health Management: Developing, Implementing, and
Maintaining a Continuous Improvement Program, *Frances Alston & Emily J. Millikin*

Handbook of Emergency Response: A Human Factors and Systems Engineering Approach,
Adedeji B. Badiru & LeeAnn Racz

Handbook of Industrial Engineering Equations, Formulas, and Calculations, *Adedeji B. Badiru &
Olufemi A. Omitaomu*

Handbook of Industrial and Systems Engineering, Second Edition, *Adedeji B. Badiru*

Handbook of Military Industrial Engineering, *Adedeji B. Badiru & Marlin U. Thomas*

Industrial Control Systems: Mathematical and Statistical Models and Techniques,
Adedeji B. Badiru, Oye Ibidapo-Obe, & Babatunde J. Ayeni

Industrial Project Management: Concepts, Tools, and Techniques, *Adedeji B. Badiru,
Abidemi Badiru, & Adetokunboh Badiru*

Inventory Management: Non-Classical Views, *Mohamad Y. Jaber*

Kansei Engineering—2-volume set
- Innovations of Kansei Engineering, *Mitsuo Nagamachi & Anitawati Mohd Lokman*
- Kansei/Affective Engineering, *Mitsuo Nagamachi*

Kansei Innovation: Practical Design Applications for Product and Service Development,
Mitsuo Nagamachi & Anitawati Mohd Lokman

Knowledge Discovery from Sensor Data, *Auroop R. Ganguly, João Gama, Olufemi A. Omitaomu,
Mohamed Medhat Gaber, & Ranga Raju Vatsavai*

Learning Curves: Theory, Models, and Applications, *Mohamad Y. Jaber*

Managing Projects as Investments: Earned Value to Business Value, *Stephen A. Devaux*

Modern Construction: Lean Project Delivery and Integrated Practices, *Lincoln Harding Forbes &
Syed M. Ahmed*

Moving from Project Management to Project Leadership: A Practical Guide to Leading Groups,
R. Camper Bull

PROJECT MANAGEMENT FOR RESEARCH

A Guide for Graduate Students

Adedeji B. Badiru

Christina F. Rusnock

Vhance V. Valencia

CRC Press
Taylor & Francis Group
Boca Raton London New York

CRC Press is an imprint of the
Taylor & Francis Group, an **informa** business

CRC Press
Taylor & Francis Group
6000 Broken Sound Parkway NW, Suite 300
Boca Raton, FL 33487-2742

© 2016 by Taylor & Francis Group, LLC
CRC Press is an imprint of Taylor & Francis Group, an Informa business

No claim to original U.S. Government works

ISBN 13: 978-1-4822-9911-3 (hbk)

Library of Congress Cataloging-in-Publication Data

Names: Badiru, Adedeji Bodunde, 1952-, author. | Rusnock, Christina F., editor. | Valencia, Vhance V., editor.
Title: Project management for research : a guide for graduate students / Adedeji B. Badiru, Christina F. Rusnock, and Vhance V. Valencia.
Description: Boca Raton, FL : CRC Press/Taylor & Francis Group, [2016] | Series: Industrial Innovation Series ; 42 | Includes bibliographical references and index.
Identifiers: LCCN 2015037049 | ISBN 9781482299113 (alk. paper)
Subjects: LCSH: Academic writing--Study and teaching (Higher) | Project management--Study and teaching (Higher) | Project management--Research. | Project management--Planning. | Technical writing--Study and teaching (Higher) | Communication of technical information--Study and teaching (Higher) | Communication in science--Study and teaching (Higher) | Dissertations, Academic--Authorship.
Classification: LCC P302.18 .B34 2016 | DDC 001.4--dc23
LC record available at http://lccn.loc.gov/2015037049

Visit the Taylor & Francis Web site at
http://www.taylorandfrancis.com

and the CRC Press Web site at
http://www.crcpress.com

Contents

Section II: Preplanning and exploration: What do you plan to do?

Section III: Planning: Making a schedule and getting organized!

Section IV: Project execution and control

Section V: Project phase-out: When is research complete?

Preface

Graduate research is a complicated process, which many undergraduate students aspire to undertake. The complexity of the process can lead to failures for even the most brilliant students. Success at the graduate research level requires not only a high level of intellectual ability, but also a high level of project management skills. After many years of supervising several graduate students, we have found that most graduate students have the same basic problems of planning and implementing their research. The forerunner for this book is a graduate guide first published by one of the authors in 1994. With the present team of authors, the book has been expanded in scope and contents to reflect the needs of today's graduate students. All graduate students need the same "mentoring and management" guidance that has little to do with their actual classroom performance. It is our conjecture that graduate students can do a better job of their research programs if a self-paced guide is available to them. This book provides such a guide. The book covers topics ranging from how to select an appropriate research problem to how to schedule and execute research tasks. The book takes a project management approach to planning and implementing graduate research in any discipline. The book uses a conversational tone to address the individual graduate student. It is a self-paced guide that will help graduate students and advisors to answer most of the basic questions of conducting and presenting graduate research. The authors believe the book will alleviate frustration on the part of both student and research advisor. Specific guidelines and examples are presented throughout the text and more detailed examples are presented in reader-friendly appendices at the end. If a graduate student is organized and prepared to handle basic research management functions, he or she and the advisor will have more time for actual intellectual mentoring, knowledge transfer, and a more rewarding research experience.

Adedeji B. Badiru
Christina F. Rusnock
Vhance V. Valencia

Authors

Adedeji B. Badiru is dean of the Graduate School of Engineering and Management at the Air Force Institute of Technology (AFIT). He was previously professor and head of Systems Engineering and Management at the AFIT, professor and department head of Industrial & Information Engineering at the University of Tennessee in Knoxville, and professor of Industrial Engineering and dean of University College at the University of Oklahoma, Norman. He is a registered professional engineer (PE), a certified project management professional (PMP), a fellow of the Institute of Industrial Engineers, and a fellow of the Nigerian Academy of Engineering. He earned a BS in industrial engineering, an MS in mathematics, and an MS in industrial engineering from Tennessee Technological University, and a PhD in industrial engineering from the University of Central Florida. His areas of interest include mathematical modeling, project modeling and analysis, economic analysis, systems engineering, and efficiency/productivity analysis and improvement. He is the author of over 25 books, 34 book chapters, 70 technical journal articles, 110 conference proceedings and presentations. He also has published 25 magazine articles and 20 editorials and periodicals. He is a member of several professional associations and scholastic honor societies.

He has won several awards for his teaching, research, and professional accomplishments. He is the recipient of the 2009 Dayton Affiliate Society Council Award for Outstanding Scientists and Engineers in the education category with a commendation from the 128th Senate of Ohio. He won the 2010 IIE/Joint Publishers Book-of-the-Year Award for coediting *The Handbook of Military Industrial Engineering*. He also won the 2010 ASEE John Imhoff Award for his global contributions to industrial engineering education, the 2011 Federal Employee of the Year Award in the managerial category from the International Public Management Association, Wright Patterson Air Force Base, the 2012 Distinguished Engineering Alum Award from the University of Central Florida, and the 2012 Medallion Award from the Institute of Industrial Engineers for his global contributions in the advancement of the profession. In February 2013, Professor Badiru was selected as a finalist for the Jefferson Science

Fellows (JSF) program by the U.S. National Academy of Sciences and the U.S. Department of State. He is the winner of the latest 2015 Public Service Award by the U.S. Air Force. Badiru has served as a consultant to several organizations around the world. He holds a leadership certificate from the University of Tennessee Leadership Institute. He is on the editorial and review boards of several technical journals and book publishers. Professor Badiru has served as an industrial development consultant to the United Nations Development Program. He is also a program evaluator for ABET (Accreditation Board for Engineering and Technology, Inc.).

Major Christina F. Rusnock is an assistant professor of systems engineering at the Air Force Institute of Technology (AFIT). She is currently a faculty research fellow for AFIT's Graduate School of Engineering and Management and program chair of the Systems Engineering Distance Learning Program. She is a member of the Institute of Industrial Engineers, the Human Factors and Ergonomics Society, and the Association of Military Industrial Engineers. She earned her BA in economics-government from Claremont McKenna College, an MS in research and development management from the Air Force Institute of Technology, and an MS and a PhD in industrial engineering from the University of Central Florida. Her research interests include human performance and process modeling for adaptive automation, cyber operations, and healthcare systems.

Prior to her appointment at AFIT, Major Rusnock served in a number of operational and staff tours in the U.S. Air Force. Her experience includes project management for the C-17 avionics and mission systems, where she was responsible for a $50M annual budget developing and implementing nine distinct C-17 modernization efforts for avionics, defensive systems, mission systems, and test laboratory facilities. In this capacity, she led a government and contractor team including experts from engineering, finance, logistics, test, major defense contractors, and end users. Prior to her work for the C-17 program, she served as a missile warning systems project manager, where she developed a source selection plan and source selection evaluation criteria for a $70M program research and development announcement contract. Her awards and decorations include the Meritorious Service Medal, the Air Force Commendation Medal, the Institute of Industrial Engineers Gilbreth Memorial Fellowship, Mobility Systems Wing Company Grade Officer of the Year, AFIT Mervin E. Gross Award, AFIT Distinguished Graduate, and Supplemental Officer Space 100 Training Distinguished Graduate. Apart from academia and military service, Major Rusnock enjoys yoga, healthy cooking, reading science fiction, and playing table-top strategy games.

Major Vhance V. Valencia is an assistant professor of systems engineering at the Air Force Institute of Technology (AFIT). He is currently a faculty research fellow for AFIT's Graduate School of Engineering and Management, director of the Graduate Engineering Management program, a registered professional engineer (PE), and a member of the Society of American Military Engineers (SAME). He earned his BS in mechanical engineering from San Diego State University and then pursued his graduate studies at AFIT, earning his MS in engineering management and PhD in systems engineering. His research interests include infrastructure asset management, risk to infrastructure systems, critical infrastructure protection, applications of autonomous systems for civil infrastructure, network modeling, and additive manufacturing for civil engineering systems.

Prior to his appointment at AFIT, Major Valencia served in a number of operational and staff tours in the U.S. Air Force. His experience includes construction contract management for the Air Force Civil Engineer Center in Europe where he oversaw large-scale military construction projects totaling $56M. He led a group of 29 engineers in the installation planning, development, and construction programs for Al Udeid Air Base, Qatar and, while at Langley Air Force Base, VA, he led a group of 19 emergency managers for contingency planning/disaster response, deployment readiness, and chemical and biological warfare defense training. He has served several tours overseas that include stations in Germany, Qatar, and Iraq. His awards and decorations include the Meritorious Service Medal, three awards of the Air Force Commendation Medal, the Air Force Achievement Medal, the SAME European Regional Vice President's Medal, and the SAME Leadership Award. Apart from academia and military service, Major Valencia pursues his interests in distance running, music, and the maker movement that includes DIY (do-it-yourself) fixes and projects often involving the voiding of a manufacturer's warranty.

section one

An introduction and overview to project management for research

chapter one

Defining project management for research

> The mere formulation of a problem is far more often essential than its solution, which may be merely a matter of mathematical or experimental skill. To raise new questions, new possibilities, to regard old problems from a new angle requires creative imagination and marks real advances in science.
>
> **Albert Einstein**

Introduction

In order to manage research, you must understand what constitutes research. In order to use project management (PM) to manage research, you must understand what PM entails.

Several definitions exist in the literature for research and PM. This book presents the general definitions, concepts, principles, and techniques of research execution from a PM perspective.

Coauthor Christina Rusnock defines research as an application of the scientific method, which begins with a hypothesis, proceeds to the collection of data, which requires analysis and interpretation, and concludes with an assessment of the hypothesis.

Leedy and Ormrod (2013) define research as "a systematic process of collecting, analyzing, and interpreting information—data—in order to increase our understanding of a phenomenon about which we are interested or concerned (pp. 2–3)." Further, Leedy and Ormrod (2013) suggest that research has the following eight distinct characteristics:

1. Research originates with a question or problem
2. Research requires clear articulation of a goal
3. Research usually divides the principal problem into more manageable subproblems
4. Research is guided by the specific research problem, question, or hypothesis

5. Research requires a specific plan for proceeding
6. Research rests on certain critical assumptions
7. Research requires the collection and interpretation of data in an attempt to resolve the problem that initiated the research
8. Research is, by nature, cyclical or, more exactly, helical

For the purpose of graduate research, coauthor Adedeji Badiru (Badiru, 1996) concisely defines research as "developing a new idea, experimenting with it, and proving that it works or not does work (p. 37)." This requires a very careful research formulation strategy. A feasible and compatible research topic must be selected for the research to be successful.

Based on these definitions and the embedded concepts, it is obvious that a management technique is essential for executing a research agenda. Specifically, you should treat any research undertaking as a conventional project that is amenable to the application of the conventional tools and techniques of PM.

You should address the following typical questions when initiating your graduate research:

- Does the topic fit your background and academic preparation?
- How well does the topic fit your personal and career interests?
- Is the topic likely to lead to a new contribution to the field?

A common problem in graduate research is determining what new work has been done by the student research and what has been culled from the existing literature. To address this problem, a detailed write-up must be presented on the proposed methodology of the research.

One source of motivation for a research topic is the real-world work environment. Graduate students who have been in the real-world work environment often have a good idea of what their research topic should be. Therefore, you must formulate your overall research goal in terms of achievable objectives. Each research effort faces three major constraints:

1. Performance expectations
2. Schedule requirements
3. Cost limitations

Performance expectations relate to the anticipated, envisioned, or specified outputs of the research project. The outputs of a research project can be defined in terms of products, services, or results. An example of a product is a new software development. An example of a service is a new process or a set of procedures to meet a customer's need. An example of a result is a new optimized layout for a production facility.

Schedule requirements establish the time-based constraints on the research project. You must be conversant with institutional and

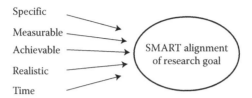

Figure 1.1 SMART alignment of research goal.

departmental timelines for accomplishing specific milestones in your research project. In addition to the externally imposed schedule requirements, you may also want to set your own personal time constraints to guide your research activities. In this regard, it is always helpful to incorporate buffers into your schedule to take care of contingencies.

Cost limitations specify the budgetary boundaries for your research project. Time and material costs are key elements of your research. You must strategically allocate your available resources to all the activities entailed in your research. You certainly don't want to run out of resources when your project is only half-way complete. Like they say, "time is money." So, you must manage your time the same way you manage your money. If you squander your time on non-value-adding activities, it is just like wasting your money on unimportant purchases. In fact, most graduate research projects will be more time-management dependent than on hard-dollar expenditures.

The SMART (specific, measurable, achievable, realistic, and timed) principle of PM is quite applicable to managing the major project constraints outlined above. Figure 1.1 illustrates using the SMART principle to properly align your research goal and we discuss this a little more in Chapter 6. Chapter 10 also uses the SMART principle in the context of managing interpersonal relationships with your advisor.

To adequately address the prevailing constraints, you must be ready for a paradigm shift, if necessary, in your research endeavors. A paradigm is a model or framework within which a problem is solved. It defines the boundaries and rules that guide the problem-solving approach. A paradigm shift requires looking at the other side of the coin, so to speak. The idea that what was successful in the past will continue to be successful does not pave the way for research success.

Types of research

The word research comes from a prefix and a root word meaning "to seek out again." Most graduate research studies do involve seeking out ideas and materials already developed or documented by others. In this process, new discoveries are made. Some research endeavors, particularly in science, focus on finding things that have never before been known or documented. In this process, new inventions may develop. There are different types of research described as follows:

Laboratory research: Research performed in a laboratory setting. This may use live test subjects, the proverbial guinea pigs, or inanimate laboratory materials.

Experimental research: Research performed by setting up carefully designed experimental procedures. It involves studying the effects of certain combinations of independent factors (treatments) on some dependent factor. Most laboratory research studies are experimental in nature.

Pure research: Research done purely for scientific interest and curiosity. It aims to add to the body of scientific knowledge. It often deals with physical and scientific objects and phenomena. It may not yield immediate or practical application potential.

Scholarly research: Research effort whereby the research deals primarily with written documents rather than physical and scientific objects, as in pure research. The publication of papers in technical journals and presentation of technical papers at professional conferences fall under the category of scholarly research.

Applied research: Research that focuses on practical applications of what has already been discovered, developed, documented, or theorized.

Theoretical research: Research that addresses the development of new concepts based on proven scientific merit. It adds to the body of knowledge (BOK) from which practical applications can be drawn later on.

Technical research: Research involving the study or development of physical objects with the intended purpose of practical utility. It may address the functionality of the object or the scientific merit of its configuration.

Business research: Research dealing with information needed to make business-oriented decisions. It often involves the development of strategic business plans.

Economic research: Research that addresses the processes of interactions of economic factors and their implications on business, social, and political events.

Market research: Research involving the study of what the market (consumers) wants. This helps to determine the type, nature, and quantity of products to develop for consumers.

It is very important to distinguish between your own type of graduate research and the different types of research described above. To avoid confusion, you should recognize that graduate research is often focused on the creation of some new knowledge, some new approach, or some new tool for solving a specific problem. In comparison, a market research by a business may focus on an assessment of the market potential to sell new products.

PM defined

Based on the foregoing presentations, it is now appropriate to provide a discussion of the basic principles of PM (Badiru, 2009). A project is traditionally defined as a unique one-of-kind endeavor with a specific goal that has a definite beginning and a definite end. A project is a temporary endeavor undertaken to create a unique product, service, or result. Temporary means having a defined beginning and a definite end. The term unique implies that the project is different from other projects in terms of characteristics. In this respect, your graduate research fits the definition of a typical project and is quite amenable to the formal application of PM techniques. Badiru (2009) defines PM as the process of managing, allocating, and timing resources to achieve a given goal in an efficient and expeditious manner. Once again, this definition directs or fits the process of executing a research project.

A PM methodology defines a process that a project team uses in executing a project, from planning through phase-out. A PM information system refers to an automated system or computer software used by the PM team as a tool for the execution of the activities contained in the PM plan. A PM system is the set of interrelated project elements whose collective output, through synergy, exceeds the sum of the individual outputs of the elements.

> *Program versus project*: A program is defined as a recurring group of interrelated projects managed in a coordinated and synergistic manner to obtain integrated results that are better than what is possible by managing the projects individually. Programs often include elements of collateral work outside the scope of the individual projects. Thus, a program is akin to having a system of systems of projects, whereby an entire enterprise might be affected. While projects have definite end points, programs often have unbounded life spans.

> *Identification of stakeholders*: Stakeholders are individuals or organizations whose interests may be positively or negatively impacted by a project. Stakeholders must be identified by the project team for every project. A common deficiency in this requirement is that organization employees are often ignored, neglected, or taken for granted as stakeholders in projects going on in the organization. As

the definition of stakeholders clearly suggests, if the interests of the employees can be positively or negatively affected by a project, then the employees must be viewed as stakeholders. All those who have a vested interest in the project are stakeholders and this might include the following:

- Customers
- Project sponsor
- Users
- Associated companies
- Community
- Project manager
- Owner
- Project team members
- Shareholders

PM knowledge areas applied to research

The traditional PM BOK compiled by the Project Management Institute (PMI) contains areas that are directly applicable to managing a research project. Although not all the elements of the PM BOK will be relevant for all research projects, we describe all the elements below for the sake of completeness. You will pick and choose whichever applies to your own research project from the areas outlined as follows:

1. Research integration
   a. Research project charter
   b. Research scope statement
   c. Research management plan
   d. Research execution management
   e. Research control
   Research integration is essential for cases where your research outputs will feed into other projects for the purpose of achieving an overall integrated system.
2. Research *scope* management
   a. Focused statement of research (SOR)
   b. Research cost/benefits analysis
   c. Research constraints
   d. Research work breakdown structure
   e. Research activity breakdown structure
   f. Research change control
   Research scope management is essential for ensuring that your research does not grow needlessly and endlessly. The fear of a graduate student is that the research advisor will add new requirements

and expectations to the research as it goes along. Research scope
management can help control ad hoc growth of your research.
3. Research *time* management
   a. Research schedule planning and control
   b. Activity network analysis using critical path method (CPM)
   c. Nondeterministic activity network analysis using program eval-
       uation and review technique (PERT)
   d. Research resource allocation
   e. Research time reporting
4. Research *cost* management
   a. Research financial assessment
   b. Research cost estimation
   c. Research output forecasting
   d. Research cost control
   e. Cost reporting
      Time and cost are the basis for research accomplishment. These
   attributes of your research must be managed to ensure a successful
   end result.
5. Research *quality* management
   a. Research total quality management
   b. Research quality assurance
   c. Research quality control
   d. Research quality cost assessment
   e. Research quality conformance
      Quality is the tangible expression of your research accomplish-
   ment. The quality of the work itself as well as the quality of the
   research report speak of your contributions and must be expressed
   explicitly.
6. Research *human* resource management
   a. Research leadership
   b. Research team management
   c. Research team building
   d. Research conflict resolution
   e. Research personnel compensation
   f. Research organizational structure
7. Research *communication* management
   a. Research communication matrix
   b. Research communication modes
   c. Research presentation
8. Research *risk* management
   a. Research risk identification
   b. Research risk analysis
   c. Research risk mitigation
   d. Research contingency planning

9. Research *procurement* management
    a. Research material selection
    b. Research vendor prequalification
    c. Research contract types
    d. Research contract risk assessment
    e. Research contract negotiation
    f. Research contract change orders

The above segments of the BOK of PM cover the range of functions associated with any research project, particularly complex ones. Interdisciplinary research, in particular, poses unique challenges pertaining to efficient communication, credible leadership support, dependable personnel interfaces, updated technology, progressive operating environment, and alignment of research interests. Assessing and resolving concerns about these issues in a step-by-step fashion will create a foundation of success for a large research project. While you may not consider your research studies to be as grand as corporate projects, the management framework for dealing with the issues remains the same. While no system can be perfect and satisfactory in all aspects (and there will be times when you consider your work less than perfect and satisfactory), a tolerable trade-off on the factors is essential for your research success.

Some of the trade-offs that you will encounter in your research journey may be in terms of time and cost. For every month that you spend in graduate school, that is a month less of income you are earning at work and another month of tuition cost that you incur. Or maybe you are trading off different "buckets" of time: study time versus socializing time versus parenting time versus research time versus time for sleep. Another typical trade-off might deal with the amount of risk you are willing to take with your research versus the quality of the work. A high-risk research project might garner you a more interesting and publishable research result versus a low-risk research project that may lead to less-than-enthusiastic praise from your research advisor. Dealing with these trade-offs is essential in understanding the many aspects of your research project and will help with increasing the likelihood of your success.

Applicability of PM to graduate research management

PM has general applicability to every human endeavor and its use continues to grow rapidly. Ancient and contemporary projects have benefited from PM practices. Records indicate that even the technology of the ancient world practiced PM. The need to develop effective management tools increases with the increasing complexity of new technologies and processes. The life cycle of a new product to be introduced into a competitive market is a good example of a complex process that must be managed with integrative PM approaches. The product will encounter management functions as it goes from one stage to another. PM will be needed throughout the design and production stages of the product. PM will be needed in developing marketing, transportation, and supply chain strategies for the product. When the product finally gets to the customer, PM will be needed to integrate its use with those of other products within the customer's organization. The need for a PM approach is established by the fact that a project will always tend to increase in size even if its scope is narrowing. An integrated PM approach can help diminish the adverse impacts of project complexity through good project planning, organizing, scheduling, and control.

PM represents an excellent basis for integrating various management techniques such as finance, economics, operations research, operations management, forecasting, quality control, queuing analysis, and simulation. Traditional approaches to PM use these techniques in a disjointed fashion, thus ignoring the potential interplay between the techniques. The need for an integrated PM worldwide is evidenced by the repeated reports from the World Bank, which acknowledges that there is an increasing trend of failed projects around the world. The bank has loaned billions of dollars to developing countries over the last half century, only to face one failed project after another. The lack of an integrated approach to managing the projects has been cited as one of the major causes of project failures. This is particularly crucial for graduate research projects, which require a systematic integration of technical, human, and financial resources to achieve organizational goals and objectives.

Project closeout and lessons learned

Mistakes are an essential part of learning. Learning is essential for future project success. Plan your project appropriately. Execute your project as planned. Learn from the project and document lessons learned, as well as best practices. It is essential to close out a project forthrightly. Use the project closeout to plan and initiate the next project. Not closing out a project properly is a sure path to failure in subsequent related projects.

We discuss dealing with mistakes and failure and project closeout in Chapters 9 through 11 of this book. Graduate research can have many pitfalls that are difficult to recover from if mistakes or plans gone awry are not dealt with quickly and appropriately. Although we hope that you achieve every success in your graduate research, we also hope that you are armed with the tools needed in facing challenges that you'll come across in your program.

Conclusion

Since academic research endeavors are complex projects, they are a natural fit for the application of formal PM. This book provides graduate students as well as other researchers with a general guidance for conducting academic research as well as specific tools for research execution. Recognizing the category of products expected from your research is an essential part of managing your research more effectively. All project outputs can be categorized into levels and types. This makes the application of the PM process applicable to every undertaking because each effort is expected to generate an output in one or more of the following categories:

- *Product*: Physical products (e.g., thesis document, technical report)
- *Service*: Business process (e.g., new experimental procedure)
- *Result*: Knowledge creation (e.g., the act of research, your education, and training)

The basic sequence of executing a research project is summarized below:

Concept—Planning–organizing–scheduling–project execution–control–phase-out

In this book, we sequence the contents to follow the flow above. In section one, the general concepts of PM are covered. In section two, planning is the main focus. A plan is the map for actionable items. As such, sufficient effort must be dedicated to the planning stage of your research. Section three covers organizing and scheduling. A schedule is the application of resources to a timeline. A schedule entails a sequential location of milestones along the timeline. With this, your research progress can be monitored. Section four presents guidance for research execution and control. Based on the output of the monitoring action, specific control actions can be determined to keep the research on track. Section five, the last segment of the book, covers the important requirement of phase-out. A research project should be terminated and phased out when it has reached a logical conclusion.

By following the managerial framework provided in this book, your graduate research projects can be better managed and executed successfully for the benefit of all stakeholders, including yourself.

References

Badiru, A. B. 1996, *Project Management for Research: A Guide for Engineering and Science*, Chapman & Hall, London, UK.

Badiru, A. B. 2009, *STEP Project Management: Guide for Science, Technology, and Engineering Projects*, Taylor & Francis, CRC Press, Boca Raton, FL.

Leedy, P. D. and Jeanne, E. O. 2013, *Practical Research: Planning and Design*, 10th edition, Pearson Education Limited, Essex, UK.

chapter two

Personal aspects of graduate education

This chapter takes a look at how graduate education differs from undergraduate education, the decision of going to graduate school, selecting a program, and selecting a school. If you have already started your graduate education, be sure to read at least the first section of this chapter.

What is graduate education?

Grad school is just undergrad version 2.0, right?

Your K–12 education, and even your undergraduate education, was quite likely built upon a model of the teacher who presented you with information that you consumed, digested, and then repeated back in the form of tests, papers, or projects. As you progressed, the material became more complex, dense, and plentiful, but the basic premise of the student as consumer remained largely the same. Graduate education, however, is different. In graduate education, the main purpose is no longer to be a consumer of already-established information, but rather to become a producer of new knowledge.

In order to be a producer of new knowledge, the primary focus of the educational experience shifts from teaching to research. Thus, the objectives and course material in graduate education are often less defined and the process is more self-driven. You will have more control and responsibility over your education, both in terms of what you learn and the type of products that you produce. Your previous education was largely driven by your teachers, whereas your graduate education will be largely driven by you. Therefore, the amount of effort you put into it will directly correspond with what you get out of it.

Graduate education focuses on results more than effort. The course work and course grades will depend on the quality of a small number of products, rather than being comprised of numerous assignments designed to reward effort. With graduate education, your goal is to become a specialist in a very specific field, unlike K–12 and even undergraduate education, which often seeks to make you a generalist, familiar with numerous fields.

Graduate education is also a time for networking. It is where you begin to build your professional and academic reputation. The other graduate students and even the faculty will soon be your peers and coauthors; thus, it is important to conduct your graduate education in a professional manner and to consider this part of your professional journey, rather than the end of your educational journey.

So, the thesis/dissertation is just a big paper, right?

While your master's thesis or PhD dissertation could be the longest document you ever write, it is not simply a longer version of the class papers you may have written in the past. With a class "research" paper, you actually did little to no actual research. Research requires an investigation, data collection, data analysis, and synthesis. Class papers typically involve looking up information that other academics have already established and then compiling this information into a single document. In order to qualify as research, the author must make an original contribution to the body of knowledge. This requires being familiar with the current state of the body of knowledge, which can be a never-ending, and very daunting task. Once the researcher has established the current state of the body of knowledge, they can identify "gaps" in this knowledge.

A master's thesis and a PhD dissertation have many similarities. The thesis and dissertation both require identifying a gap, creating a plan to resolve that gap, and executing a study that will fill the gap. The thesis will likely make a single, small contribution; whereas the dissertation will make a larger, and often several, contribution(s). The thesis typically has one major hurdle: the defense. The dissertation often includes several hurdles: qualifying exam, candidacy, proposal, and a defense. The types and nature of these hurdles differ among institutions, but the dissertation process is designed to be more robust and to produce numerous publications. For more details on the products created from the graduate process, see Chapter 11.

Is graduate education the right fit for you?

Graduate education is not right for everyone. Graduate education requires a large investment of both time and money, so before you embark on this adventure, be sure you are choosing the right adventure.

We recommend you consider your long-term career goals and whether or not graduate education will productively contribute to them. For some career paths, trade schools and technical certifications may actually give you more appropriate skill sets and better equip you for finding a job. The decision to attend a graduate program should consider your career goals, the current job market, and your level of experience.

Career goals

Whether or not graduate education is right for you depends on your end goal. Just as all successful projects should have a well-defined end goal before they are begun, a potential student should have a well-defined end goal before they embark on graduate education. Once you know what your end goal is, you can then determine whether the graduate education will help you get there.

If you have no idea where you want to end up, consider gaining some work experience first. Internships or job-rotation programs are a great way to gain exposure to the potential career options available to you. Identifying the type of career path you want to pursue will help to assure that when it is time to go to graduate school, you get the right degree. You don't want to spend 2–5 years of your life pursuing a technical degree, only to discover later in life that you should have got a management degree (or vice versa).

Once you have identified a general career path, talk to mentors and leaders in that career field to find out what type of education they received, what the timing of that education was, and what are the current expectations for up-and-coming, young professionals in that field. You can also use your Internet search skills to examine biosketches, resumes, and curriculum vitae of successful professionals in the career field. Don't be shy about contacting professionals and human resources departments at your future potential employers to find out about the type of experience, education, and skill sets that are needed for your future career path. Most people are more than happy to give advice.

Current job market

Another consideration is the current job market. Of course, this is not the job market as a whole, but rather the job market for the specific career you have selected. Certain fields, such as law and medicine, will require advanced degrees rather early in a career. However, even these fields offer a wide range of paths and, thus, a wide range of education opportunities. For example, paralegals do not require the same education as lawyers; registered nurses do not require the same education as doctors. It is important to identify the necessary certifications and educational requirements for your specific path, and in some cases you may need to obtain these before you begin working. In other cases, advanced education can be obtained in parallel with, or even after, beginning your career. The structure of your particular job market will reveal not only the type of education you need to obtain, but will also indicate the timing of when that education should be obtained.

Employment rates go through cycles, so the current state of the economy may have some influence over the timing of your education. Since long periods of joblessness can negatively impact your future employment prospects, it may be best to time your education with periods of high unemployment. For your resume, it is better to be in school than to show a gap of unemployment. But keep in mind that school is usually not free, so weigh the cost of going into debt with the cost of being temporarily unemployed.

Level of experience

Ideally, it is best to balance work experience, educational experience, and professional development. From the employer's perspective, the best job applicants have an appropriate mix of these three aspects. A job applicant with too much education and too little experience is likely to be viewed more as a liability rather than an asset, because they lack the experience to demonstrate that they can actually handle the job, but often require pay commensurate with their degree. Thus, having a master's degree with 2–4 years of job experience can be more attractive than a PhD with no work experience.

Take a hard look at your resume and figure out where you are lacking. If you have trouble figuring out where your resume is weak, discuss your resume with a mentor or someone who has achieved what you would like to achieve 5–10 years from now. If your weakness turns out to be lack of graduate education, then it is the time to pursue a degree. If your weakness is in work experience, then build that part of your resume first. Once you have 2–5 years of experience, reevaluate because the education may then become your weakest link. Building work experience doesn't necessarily put you behind the educational curve. In some cases, such as MBAs, some of the best advanced degree programs may require 3–5 years of *relevant* work experience in order to enter the program.

What should I look for in a graduate program?

Now that you have committed yourself to a career path, and identified it is the right time to get an advance degree, how do you go about selecting a graduate program? There are a number of things to consider, including cost, scholarships/grants, proximity to your current residence, length of the program, and prestige/ranking of the program. Since you already went through a similar process when selecting your undergraduate degree, we are not going to consider those factors that you likely already have experience in considering. Instead, we are going to comment on some (new) factors, that you likely did not consider when selecting your undergraduate degree.

Create variety in your education

Variety is the spice of life, and it also creates spice to your resume. Whether you end up in academia or industry, having a somewhat cross-disciplinary education adds flexibility and can also show growth. Ideally, you should consider obtaining an advanced degree from a different school and in a different, but related field from your undergraduate degree.

If you obtained a computer science undergraduate degree from University ABC and decide to get a computer science master's degree from the same school, you will very likely encounter the same professors, with the same perspective on the discipline, teaching the same ideas you encountered in your undergraduate education. However, attending University XYZ for your master's program will expose you to new professors, perspectives, and ideas, creating a much richer learning environment for you. In addition, you are armed with the ideas and perspectives from University ABC, so you actually provide a positive contribution to University XYZ's program as well. It is tempting to attend a school that is familiar and comfortable, whereas attending a program at a new school is likely to result in increased professional and personal growth—the whole reason you are seeking a degree.

In addition to attending a different school from your undergraduate education, we also urge you to consider selecting a related, but different field. Of course, the specific degree needs to be tied to your career goals, which might dictate a specific degree, but if you do have flexibility, selecting a slightly different field will enable you to think more broadly; allow you to integrate tools, methods, and ideas across disciplines; and make you more valuable to your future employer. In addition, if you find that for personal reasons (cost, family, etc.) it is most practical for you to attend the same school as your undergraduate degree, selecting a different field can provide similar advantages of being exposed to new professors and ideas.

Industry path

If your future career path involves obtaining employment in industry, you may also want to examine the job placement rates and job placement programs that are offered by the various schools you are considering. Note that some schools attempted to "inflate" their graduate job placement rates by subsidizing internships for their first year graduates, so if possible look at not only the number of graduates that find jobs within the first year of graduation, but also their long-term employment rates.

If you have a specific future employer in mind, find out if that employer recruits new-hires from specific university programs. Also, find out what your future employer is looking for in job applicants. If there are specific gaps that the firm is trying to fill, you can make yourself an asset to the company by selecting a program that will give you the relevant skill sets. If the company is small, you may be able to talk to the hiring authority directly; perhaps even arrange to do a co-op while attending school. If the company is large, examine the hiring announcements published by the human resources department in order to understand what types of graduate education the company is looking for.

Academic path

If your future career path is an intention to enter academia, you need to make sure that the program you enter has enough academic rigor that you will be able to continue successfully in that field. Especially with PhDs, some fields (e.g., psychology) are highly apprenticeship-based. Thus, you work for and are mentored by a specific advisor, who controls and contributes greatly to your education. If you are entering this type of field, we highly recommend you consider the type of research you want to do, and identify the researchers that are highly published in that area. Then, select your school based on your desired research advisor.

For nonapprenticeship degree programs, consider universities that are well known for the specific major. A school may be well known for its biology program, but that doesn't mean that it is strong in electrical engineering. As you examine the programs of various schools, and the quality of the research they are producing in your field, you will also want to consider the opportunities for grants, fellowships, and research assistanceships. If you are looking to do research that requires specialized equipment, understanding the quality of the labs, and the types of equipment available, will also be important.

As mentioned above, variety in education is highly valued. Just as obtaining your graduate degree from the same location as your undergraduate degree limits your exposure to ideas, likewise, hiring professors that obtained degrees from the same university also causes academic stagnation. Most universities recognize this, and seek to avoid "academic in-breeding" by hiring faculty that obtained their PhD elsewhere. Thus, keep in mind that you will likely not be hired to become a professor at the university at which you obtained your terminal degree.

Which school/program do I choose?

Once you have identified the specific degree that you are seeking to obtain, your next task is to select the appropriate program to apply to. With your undergraduate program, you most likely picked a school based on factors external to the specific degree you were seeking (e.g., location, cost and school reputation); in fact, 80% of students change their major at least once and, on average, students change their major three times. However, with graduate education, you are entering a specific program, and often-times have to compete with other potential graduate students for a limited number of slots. Because you are applying to a specific program, we rec-ommend that your decision on where to apply should largely account for program-specific considerations. Of course, other considerations, such as location and cost, may need to factor into your decision as well.

When examining specific programs, seek to understand where that program ranks among its peers, and seek to discover the specific areas that the program specializes in. For example, if you are seeking a micro-biology degree, take the time to understand which areas of microbiology the program is strong in (e.g., virology, bacteriology, mycology, parasi-tology). If your interest is in bacteriology, then you should seek out pro-grams that specialize in this subdiscipline. To find strong programs in a subdiscipline, seek out published research in that area and identify the programs that those researchers belong to. You can also look at the faculty profiles on the program's website. If you are seeking a PhD in business, and would like to focus on finance, but find that 70% of the faculty in a program specializes in organizational management, then this would be an indication that the finance program is not as strong at this institution. In addition to the program specializations, you may also find it valuable to identify graduation rates, average time to obtain a graduate degree, and the job placement rates.

Once you have narrowed your search down to a manageable number of programs, we highly recommend that you take the time to do a site visit. During this site visit, you should meet with new and tenured profes-sors, and if possible other graduate students. This will help you to further understand the programs strengths and weakness, organizational cul-ture, relationship between graduate students and faculty, funding oppor-tunities, and resources/equipment that will be available to you. If you know what you are trying to achieve through your graduate education, you can identify your potential advisor prior to your visit, by examining their curriculum vitae, research interests, and publications, all of which should be available on the faculty profile of the university's website. If you have background knowledge of the professor's research, then you will be better equipped to build a rapport with this professor during your site visit. By being already familiar with the professor's research, you demon-strate that you are a good fit for the program, and can build a champion on the inside.

If you have the opportunity to speak to other graduate students during your site visit, take the opportunity to ask them about the quality of mentoring from faculty and senior students, academic and financial support provided by the department, quality of research guidance, stress and organizational culture, insight into student–faculty relationships, and things they wish they had known about the program.

A site visit is also an opportunity for you to establish realistic expectations about your chance of being accepted. Some program specializations may only accept new students every other year, due to the timing of the classes offered. Furthermore, some program specializations may only accept new students if the faculty member has research funds to support a new student (hired as a research assistant). Other programs may use department funds to pay for new graduate students as teaching assistants, but are still limited by the funds available as to how many students will be accepted.

Ideally, by the end of your site visit, you will have discovered which program(s) is a good fit, you will have identified a potential research advisor, and will have clear expectations about the program's willingness and ability to fund you as a student. If your site visits reveal that you are not a desirable candidate, find out what aspects of your education or experience are your weaknesses, and seek to resolve these, so that you can earn your place in the program you want to attend. In undergraduate education, you very likely went to a university directly after high school. Thus, the fixed application schedule incentivized you to apply to numerous schools, because you knew you had to go "somewhere." With graduate school, you have more flexibility with the timing, so it is better to attend when you are truly prepared and are able to get into the right program.

section two

Preplanning and exploration:
What do you plan to do?

chapter three

Choosing your research topic

How do I find a research topic?

Selecting your research topic can be a daunting task. If this is your first exposure to conducting independent research, you may feel like you don't have any ideas, and you may not even be sure what a good research idea even looks like. If this is the case, then, a good first step would be to provide yourself with exposure to academic research in your discipline. To gain this exposure, use your university's library to find academic journals in your field; most academic journals will have your field's name in their title. For example, *Applied Cognitive Psychology*, *International Journal of Industrial Engineering*, *Harvard Business Review*, and *Chronicle of Higher Education*. If you have trouble finding relevant journals, ask professors at your university to recommend relevant journals to you.

In order to gain wide exposure, read the titles and abstracts of recently published articles. This will give you a good feel for the types of research projects that are being conducted in your field, as well as exposure to the size, scope, and precision of research questions. While there is some variability, a conference paper or journal article is approximately the size and scope of a solid master's thesis. A PhD dissertation, on the other hand, will likely be composed of 3–4 conference papers and journal articles.

Previewing current academic literature can also help you explore your personal interests, in order to find a topic area that you are interested in. In addition to your personal interests, it is wise to consider the interests within your profession/industry and the current research that is being conducted by your department. Topics that are attracting attention in your profession/industry are likely to have an easier time-finding funding/sponsorship, which may be critical for conducting your research. Topics that are of interest to your profession are also likely to have more avenues for publication. Topics that build upon active research streams in your department will give you access to quality mentors, student collaborations, equipment, and funding. Ideally, you should try to find a topic area in the intersection of personal, professional, and departmental interests (Figure 3.1). For example, you might have a personal interest in cars and automation, your department faculty may have an interest in human–computer interfaces, and your profession/industry may have a growing interest in bringing user-friendly driverless technologies to the

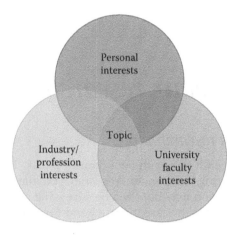

Figure 3.1 Intersection of interests.

market. These interests could be combined by conducting research on human–computer interfaces for driverless cars.

Such an endeavor may seem overly risky. After all, what is the likelihood of finding a topic that crosses all three areas of interest? First, choosing the right program is a start (go back and read Chapter 2!). Beyond that, having the courage to discuss your ideas with the faculty and your classmates will help in generating ideas both in finding the topic and in conducting your literature reviews. You'll quickly find that others might share in your passion, the department has hidden resources in the form of a different project that can be adapted from past research, or you may be pointed to the faculty outside of your current department. Being bold in your search for a research topic will create opportunities for interesting and innovative research.

How do I identify a research gap?

As discussed in Chapter 1, research is an application of the scientific method, which begins with a hypothesis, proceeds to the collection of data, which requires analysis and interpretation, and concludes with an assessment of the hypothesis. Thus, the product of a research effort is the expansion of the current BOK. In order to expand the current BOK, you must first know the current state of the BOK. Once you are familiar with the current state, you will then be able to identify the "gaps"—those areas in the BOK that are unknown. A good research project should seek to fill-in a current gap in the knowledge.

You already started to get a sense of the BOK when you previewed recent journal articles in your quest for a topic area. Once you have a topic narrowed down, it is now time to go back to the academic literature in order to delve more deeply. If available, we encourage you to begin with literature review articles. These articles are specifically designed to provide a broad overview of the current state of the literature, and they often indicate emerging trends, upcoming/future research, and current gaps. If you are unable to find a literature review article, then, carefully review the literature review sections of standard journal articles that are closely related to your topic area. Use the bibliography of highly relevant journal articles and review articles to find additional relevant articles. Use electronic search engines to conduct "forward-reviews" to find more recent articles that cite the relevant works you have already found. As you continue to read, you will find the literature review sections becoming more repetitive as you become more familiar with the state of knowledge in your topic area. When you are no longer finding new citations, but rather all articles point back to articles you have already found, then, you know you have reached the boundary of your field.

If you have a good grasp of the current state of the field, but are still struggling to identify a research gap, use the future work, recommendations, and conclusion sections of relevant journal articles, theses, and dissertations to see what gaps and limitations have already been identified by previous researchers. Use these gaps as a springboard for discussions with your thesis advisor/potential advisors.

Additional considerations when selecting a research topic

As you narrow in on potential gaps that your research project could seek to resolve, there are some additional considerations that you should keep in mind. The first consideration is funding. Will the research require a monetary investment (for equipment, lab technicians, etc.) in order to properly conduct the research? If so, how much will be needed, and does your research advisor have the ability to fund the project? Do you and your research advisor need to apply for a research grant? Are there grants available for your topic? What will you do if the grant proposal is not funded? Funding considerations can help shape your research by eliminating some alternatives and building a stronger case for others.

The next consideration is whether or not the topic will require access to specific people, companies, events, or equipment. If you are looking to do a study involving Fortune 500 CEOs, Chinese factory workers, natural disasters, or jet engines, then, do you and/or your research advisor have access to these resources or the relevant data? If not, then your topic might be brilliant, but infeasible.

Another consideration is whether or not your research involves human subjects. If so, you not only need to consider whether or not you have access to the specific population, but also need to review the requirements for conducting research on human subjects, and familiarize yourself with your institution's Institutional Review Board (IRB) process. Access to subjects, permissible research activities, and time constraints with the IRB approval process could impact your ability to complete the study.

A final consideration is your timeline. We'll discuss developing a research schedule in Chapter 6, but for now, consider when you need to make a decision on your topic and the amount of time you are willing to commit to graduate school. The two go hand in hand: the longer you are willing to spend as a graduate student, the more time you have in making your choice. Or the longer you take to make your choice, the more time you'll spend as a graduate student. Chances are you don't want to be a student forever; so, you should make a timely decision on your research topic.

Defining your research question: What is the right scope?

Once you have established your research topic, you will need to scope this down into a research question. The research question should be precise and well defined, and you and your advisor should be confident that you can satisfactorily answer the question, given your allotted time frame and access to resources, including data and participants.

While research questions can come in many shapes, sizes, and formats, a well-defined research question should specify three aspects of your research: (1) the dependent variable(s), (2) the independent variable(s), and (3) the context. Excluding one or more of these aspects could make your research question unnecessarily broad or ambiguous. For example, a broad, ambiguous research question might be: How do children learn? This question points to the dependent variable (learning) but does not provide insight into the independent variable(s) or context. Because this question is so broad, it is very difficult to even quantify what this study would look like or the resources that you would need to answer this question. This question can be made considerably more precise; for example, do third graders learn multiplication more accurately through memorization of tables or repeated addition? This research question specifies the dependent variable (multiplication accuracy), the independent variable (learning method—memorization or repeated addition), and the context (third graders). With this more well-defined question, you and your advisor can envision what resources you will need (third graders who have not yet learned multiplication), how you would execute the study (divide into two groups and teach a different method to each group), and how you will measure the results (multiplication test—perhaps even pre- and posttesting). Not only does a well-defined question make it clear as to how it will be answered, but it also provides the insight needed to understand whether or not you will have access to the resources (including time) needed to satisfactorily answer the question. For a deeper discussion on creating a research question, see Chapter 5.

How can I make the most of my coursework?

As discussed in Chapter 2, graduate education differs from undergraduate education by shifting the educational experience from learning an established curriculum to performing research in order to contribute to the BOK. Thus, your research should be the center of your graduate education. Ideally, your research should not only be the focus of your thesis/ dissertation hours, but should also be your focus while you are proceeding with your coursework. To truly make the most of your graduate education, we advise that you deliberately seek to apply your coursework to your research.

The first step in applying your coursework is being very deliberate in your course selection. This is not an undergraduate education; thus, your classes should not be selected based on irrelevant motivations such as which professor grades the easiest, which class has the least homework, or what time of the day the course is offered. You will likely have a core curriculum of must-take classes, along with optional track/sequence courses and electives. You should begin by making sure that you understand the process (for PhDs) and become familiar with the timing of course offerings. Some courses may only be offered once a year or even once every other year. Being aware of these rare offerings, especially for courses required as part of your core or sequence, is vital to building a schedule that lets you progress in a timely fashion. You should also pay close attention to courses that have prerequisites and be sure to time these courses so that you will have accomplished the prerequisites.

As to the PhD process, every university is different, but often, this process consists of several milestones—qualifying exam, candidacy, proposal, and defense. Oftentimes, you will not need to complete all coursework before meeting the first milestone, the qualifying exam. However, you will likely be required to have accomplished a particular set of classes (either the core or your specialty sequence) in order to take the qualifying exam. To aid in your progression, you should seek to take those courses required for the qualifying exam as early as possible. Owing to the infrequent offerings of some courses, not taking a particular course right away could delay your ability to take the qualifying exam by a year or more. Thus, it is advisable to build your schedule around the process and rare, but required courses, with the intent of taking electives later in your coursework.

Elective courses should not be treated lightly. They are the courses that permit you to tailor your education, and thus are an important part of building the skill sets necessary for your particular research endeavor. These courses should not be selected on a whim or at random, but rather be specifically designed to advance your thesis/dissertation. By saving your electives for later in your coursework, you will be better prepared to identify the courses that will help you most in your research project.

It should also be noted that most universities permit students to take one or more elective classes in the form of an independent study. This can be done with your (potential) advisor or a domain expert in the area you wish to focus on. Independent studies are an ideal method to prepare you with the background you need to conduct your research, and may build upon a topic that is highlighted in a core/sequence course or may cover a topic that is not taught in any course at your university. An independent study should be independent, which means the student should take on the responsibility of establishing the objectives and the curriculum, with the instructor available to provide mentoring and feedback.

If you wish to create an independent study and the topic is foundational, begin by conducting an Internet search for sample syllabi for that topic. After you have found several syllabi, conduct an Internet search for textbooks on that topic. Finally, use the sample syllabi and textbooks to generate a personalized reading and assignment schedule.

If the topic is specialized, then, an Internet search for sample syllabi and textbooks may be unfruitful. If this is the case, conduct an Internet search for books (nontextbooks) and journal articles on that topic, and create a personalized reading and assignment schedule using these materials instead.

Finally, provide this proposed reading and assignment schedule to your potential instructor for an independent study for them to review and revise. Taking the initiative to identify the topics and creating a plan for your independent study will help give the potential instructor the confidence that you are capable of effectively performing an independent study.

chapter four

Choosing your advisor and committee

Success in any worthwhile project requires a cohesive team of committed people. In your research, two distinct sets of people will play a huge role in the success of your eventual thesis. This chapter discusses how to select the right advisor (who will serve as a guide and mentor throughout your research journey) and committee members (who will provide essential feedback, assistance, and critical evaluation) later in the research process. Spending time to evaluate who will play these major roles in your research, and the follow-up in asking for their participation, will be time well spent ensuring your success.

How do I select a research advisor?

Selecting the right research advisor is as important as selecting the right research topic. Your research advisor will be your partner on this research journey over the next year (master's thesis) or several years (PhD dissertation). For most students, the selection of a research advisor will come before they identify a research question, as the advisor will likely guide them in selecting and scoping an appropriate research question.

Begin the process by creating a list of professors in your department who specialize in your field. For each of these professors, find out the primary focus area(s) of their research. To do this, you can begin by examining their curriculum vitae (CVs), which will most likely be available on their faculty profile page of the university's website. Some faculty members also have personal websites or will have profiles with scholarly social networks such as Google Scholar or ResearchGate. The CVs will provide you with a list of each professor's published journal articles, book chapters, and conference papers. Pay close attention to the most recently published works, as these are the most indicative of the professor's current line of research. The CVs or your department's website may also list the titles of recently graduated students' theses and dissertations, which can help you understand the type of work being done by the students who work with that particular professor. By comparing the focus areas of each professor with your personal research interests, this investigation should help you in narrowing down the pool of potential advisors.

Once you have a pool of potential advisors, you should set up a meeting with each of them to discuss their research areas and opportunities to work with them. Arrive at the meeting fairly familiar with the professor's research, and how that intersects with your research interests, so that you are able to have a productive conversation. You should seek to gain a number of insights from this meeting:

1. How healthy is the professor's research stream in the areas you are interested in? Is this an ongoing research stream or is it at the end of the line?
2. Does the professor have funding or the ability to seek funding to support you and your research?
3. Is the professor taking on new students?
4. Do you have a good rapport with the professor? Is this someone that you would enjoy working with? Do you feel like you can discuss issues freely with this professor?
5. What is his/her work style? Is he/she very hands-on and actively engaged in the students projects? Or is he/she hands-off, providing the students with a lot of autonomy? Is this work style compatible with the environment that will best allow you to succeed?
6. Is the professor's availability, funding stream, access to resources, work style, etc. consistent with any timelines that you might have for completing your research?
7. Will this professor be able to satisfy any other objectives that you might have, for example, mentoring or networking?
8. What are the professor's expectations for his/her students with respect to research assistance, teaching assistance, publishing, oversight, etc.?

Upon conclusion of your meetings with your pool of potential advisors, reflect upon your answers to the questions above. Most likely, it will be clear to you which professor would be the most suitable advisor. If there are still several candidates, it may help to prioritize the considerations above. In the end, you want to be sure that you end up in a research situation that will be successful and productive.

If there are no suitable candidates, you may want to consider whether or not you have chosen the right program/area of study. If you are too far along in the process of your education to do a course correction, you may have to settle for a less-than-ideal advisor. Again, visit these eight considerations and figure out which one(s) is the highest priority for you. You may not find someone that meets all of your criteria, but hopefully you can find a satisfactory option.

How do I select the committee members?

The second group of people essential to your success and productivity will be your research committee members. Once your research is underway, it will also be necessary to establish the committee for your thesis or dissertation. Be sure to understand your school and department's requirements for the committee's makeup. Programs will have variations on committee membership criteria, including both numbers of committee members and source (internal vs. external). Your advisor may have preferences on your committee makeup. Knowing these general rules is a starting point for your committee.

　　The right time to start identifying your committee members is after you have selected your research topic. This ensures that you are far enough along in the process to select relevant committee members, while still early enough to benefit from their feedback.

　　While every committee is different, the primary responsibility of the committee is to uphold the academic integrity and rigor of the degree program by ensuring that the research is of high enough quality to satisfy the requirements of the degree. Some committee members may be involved during the project, but oftentimes committee members are only involved during major milestones such as candidacy (PhD), proposal (PhD), and defense (masters and PhD).

　　If your advisor is hands-on, he/she may select or suggest committee members for you. If you are given flexibility in selecting your committee's makeup, then the first step is to find out your institution's requirements regarding the minimum number of committee members and whether one (or more) of the members must be outside of the department. It is advisable to keep the committee as small as possible by keeping the size to the minimum required. Increasing the number of committee members increases the likelihood for disagreements and increases the number of expected revision items from each milestone.

　　Ideally, committee members should have differing areas of expertise, with each committee member providing a unique contribution/perspective to the research problem and with each being the single authority on matters within their domain. It is a good idea to clearly communicate to the committee member the particular area in which you see their contribution (e.g., background of the field, methodology execution, statistical analysis). If each committee member is recognized as the single authority for his/her contribution area, there is less room for disagreements between the committee members.

　　When considering potential committee members, get advice from your advisor on which faculty are domain experts in the various aspects of your research. Also, inquire about the internal dynamics that your advisor is aware of from his/her previous committee experience. Being able to identify known personality conflicts or "troublemakers" can help you to avoid a problematic committee composition.

Finally, when you have identified potential committee members, you should meet with them to discuss your research. During this meeting, try to gauge their level of familiarity, understanding, and interest in your research topic. Determine whether they will be able and willing to provide a meaningful contribution, if needed. Also determine their willingness to help/advise you if needed, and whether or not you are comfortable asking them for assistance. In the end, it is your committee that will decide whether you pass or fail your defense; thus, be sure to do all that you can to set yourself up for success.

chapter five

Research question

Equally critical to choosing the right advisor is choosing the right research question. A good research question will set you up for a successful project, and will clarify the feasibility, timeline, and required resources for your research project. This chapter discusses the key components of a research question and how to write one effectively.

What is a research question?

As you may recall from Chapter 1, research is an application of the scientific method, which begins with a hypothesis, proceeds to the collection of data, which then require analysis and interpretation, and finally concludes with an assessment of the hypothesis. A good research question is specific, substantive, and unbiased. In order to ensure that the research question is well scoped, the research question should be specific enough that one can easily identify the dependent variable(s), independent variable(s), and the context. While there is no formula that defines a research question, a good rule of thumb for a research question is that it could be reduced to a generic format of: How do/does the independent variable(s) impact the dependent variable for a specific context/population? Note the use of "how," which creates a question of substantive depth; without the "how," this question could be answered by a simple "yes" or "no."

As discussed in Chapter 3, identifying the dependent variable(s), independent variable(s), and context not only eliminates ambiguity but also provides a clear vision for what resources will be required, how to execute the study, and how to measure the results. Let's take a look at some ambiguous research questions, and compare them to ones that are well scoped.

EXAMPLE 5.1

Ambiguous: What are the effects of lack of sleep? Note that this question indicates what the independent variable is (hours of sleep), but gives us no clue to the dependent variable or context. This question could thus be answered in an infinite number of ways. For example, this could be a study on how medical residents perform at the beginning versus the end of their 30 h shifts. Or this could be a longitudinal study of how hours of nightly sleep during elementary school years affect adult height. By not including the dependent variable and context, we don't know how to perform the study, and potentially would require many different studies in order to fully answer the question.

Defined: What is the impact of sleeplessness on driving performance for teenage drivers? We have now scoped this question to include a dependent variable (driving performance) and a context (teenage drivers). By adding in this information, we can now visualize who the subject population will be, what task we will ask them to perform (driving), and we can begin to brainstorm on how we might measure this (number of times cross over yellow lines, number of cones hit, braking reaction time).

EXAMPLE 5.2

Ambiguous: How can cognitive function be improved? What is missing from this question? Can you begin to see how missing one or more of the three parts—dependent variable, independent variable, and context—results in a research question that could be answered in many ways, with each individual answer not able to fully provide a satisfactory answer? There could be lots of ways to improve cognitive function—diet, exercise, sleep, mental games, reading, work-life balance, safety, vitamin supplements, novel experiences…the list goes on and on. Will you test all of these? Of course not. Instead this question should be appropriately scoped by adding in the independent variable(s) and context.

Defined: How does playing a first-person shooter game daily affect cognitive function for elderly persons? Now we can start to see a well-defined study that can be satisfactorily answered. We have a population—elderly people, which still needs to be fully defined (55 + , 65 + , 75 + ?). We have our independent variable, first-person shooter games, which we might want to contrast with other games, and even a control group. And we know what we are interested in performing a cognitive test, most likely pre- and posttreatment. Of course, this will also need to be scoped a bit further (math skills, spatial skills, reaction time?).

EXAMPLE 5.3

Ambiguous: Does winning the World Cup matter? This question indicates the independent variable (World Cup outcome) but does not indicate a dependent variable or context. Does the outcome matter to whom, and in what way? To the retirement age of the coach? To the long-term psychological states of the players? To the future revenues of the team's sponsors and advertisers? It is not clear what type of data we are going to need—it could be anything from conducting interviews with actual players or obtaining access to the team's financial records.

Defined: How does the outcome of the World Cup impact the economy of the hosting country? This question now clearly defines not only the independent variable (win/loss outcome), but also the dependent variable (the economy) and the context (host country). We can now see that we are likely going to need macroeconomic data for the host country. Perhaps we will compare quarterly figures (e.g., gross domestic product [GDP]) both before and after the World Cup for two groups: host country wins and host country losses.

Keep in mind that research entails scientific inquiry. You are seeking to understand and measure the world as it is. You are seeking to find evidence that will support or refute your hypothesis. You are not setting out to "prove" anything, and you want to be cautious that you don't allow your personal biases, or suspicions, to affect your research. You should begin your inquiry with an open mind, and concede that you may find evidence that fails to support your hypothesis—which could be a significant finding. Beginning this endeavor with an open mind begins with an open-minded research question.

EXAMPLE 5.4

Biased: How does cockpit automation hurt pilot situation awareness during in-flight transit? This question presumes that automation has negative impacts on the pilot's situation awareness, and seeks to find the ways that situation awareness is hindered. Perhaps the research presumes that by automating tasks, the pilot is less involved with the flying task, and thus does not know what is going on. This question does not acknowledge that the automation could actually enhance the pilot's situation awareness. Perhaps by freeing the pilot from mundane tasks, the automation provides the pilot with more time to keep track of his surroundings. Or perhaps the automation includes alerts that directly contribute to greater situation awareness.

Unbiased: How does cockpit automation affect pilot situation awareness during in-flight transit? By changing "hurt" to "affect," the research question is now open-minded and capable of finding ways that the pilot's situation awareness is both helped and hindered.

Operationalizing variables

Once you have a well-defined research question, with dependent variable(s), independent variable(s), and a specified context, the next step is to examine your variables to determine if they are directly measurable. Categorize your variables as either objective or subjective measures. Objective measures are concrete and easily measured. Examples of objective measures include age, height, education, race, income, reaction time, etc. Subjective measures are abstract and difficult to measure. Examples include: fatigue, stress, health, situation awareness, and performance.

Table 5.1 Operationalizing variables

Variable of interest	Example of operationalized variable
Fatigue	Hours of sleep
	Heart rate
	Blink rate
Driving performance	Number of traffic violations
	Number of times cross over yellow lines
	Number of cones hit
	Reaction time
Cognitive function	Score on math test
	Score on spatial test
	Reaction time
Economic impact	Change in stock market index
	Change in quarterly GDP
	Change in unemployment rate

If your variables include subjective measures (and the interesting questions often do), you will need to operationalize your variable, so that you can conduct your study. Operationalization is the process of identifying a measurable, quantifiable, and valid index for a variable. Table 5.1 provides examples of a few subjective measures and potential indices that could be used to operationalize these measures. Note that in many cases, you might need several operationalized measures to truly capture the entirety of the concept encompassed by your subjective variable of interest.

What are investigative questions?

Research questions—even those that are well scoped—can often consist of multiple facets. When a research question has multiple aspects to it, it is wise to divide the research question into investigative questions (also known as subquestions). These investigative questions should consist of all of the direct questions that need to be answered in order to satisfactorily answer the research question. A well-defined investigative question should be connected to a single hypothesis and should be answerable through a single hypothesis test.

Investigative questions frequently originate in the process of operationalizing variables. For example, suppose the research question is *How does the outcome of the World Cup finals impact the economy of the hosting country?* and you have operationalized economic impact as the change in the home countries stock market index, change in quarterly GDP, and change in the unemployment rate. In this case, the dependent variable has three components; in other words, there are three (related) dependent variables. This provides the opportunity for three investigative questions:

1. What is the impact of the World Cup finals on the *stock market* of the hosting country?
2. What is the impact of the World Cup finals on the *quarterly GDP* of the hosting country?
3. What is the impact of the World Cup finals on the *unemployment rate* of the hosting country?

These three investigative questions now allow for the development of hypothesis statements that can be tested using t-tests, analysis of variance (ANOVAs), or other suitable statistical techniques. Example hypothesis statements could be:

1. Closing daily value of the stock market index will increase compared to the previous day if the host country wins the World Cup.
2. Quarterly GDP will be higher than the previous quarter for the host country of the World Cup regardless of who wins.
3. Hosting the World Cup will have no impact on annual unemployment rate.

Individual investigative questions can also result from research projects that have more than one dependent or independent variable. If there are multiple independent variables, there may also be investigative questions regarding the interactions between these variables, in addition to the main effects.

The investigative questions, along with their associated hypothesis, should reveal what experiments or data collection will be required and which statistical methods will be appropriate to answer the question. A manageable thesis or dissertation will likely have between three and five investigative questions. If the research question requires significantly more than five questions in order to be satisfactorily answered, then the research question is most likely too broad in scope or too ambiguous. If this is the case, efforts should be made to reduce the scope and define the question more precisely.

What are pseudo-investigative questions?

Pseudo-investigative questions are questions that the researcher needs to answer in order to answer the research question, but do not require an actual research investigation in order to answer. Pseudo-investigative questions typically arise from the researcher's lack of familiarity with the subject matter or current state of the body of knowledge. They are real questions that need to be answered in order for the research to progress, but they have likely already been answered by other researchers, and can typically be resolved through a literature review. For example, using the research question *How does cockpit automation affect pilot situation awareness during in-flight transit?*, a researcher may have the following pseudo-investigative questions:

- What aspects of the cockpit can be automated?
- How is situation awareness measured?
- What events occur during flight transit?
- Is there a simulator or synthetic task environment that I can use to conduct experiments?

These pseudo-investigative questions ask questions that can be answered from literature reviews, discussions with subject-matter experts, direct observation, or consultation with department faculty. The answers to pseudo-investigative questions often provide the background and insights required to develop investigative questions, operationalize variables, and construct experiments and experimental design. Thus, they are a necessary step, but they are not true investigative questions. They can be differentiated from investigative questions, because pseudo-investigative questions typically don't have associated hypotheses, don't require experimentation, and don't require analyses and interpretation of collected data.

Using investigative questions and pseudo-investigative questions to create a research plan

After identifying your research question, investigative questions, and pseudo-investigative questions, you can begin to construct your research plan. Chapter 6 details numerous techniques for creating and scheduling your research plan. In this section, we will look at how to generate some of the content for that research plan by using the investigative and pseudo-investigative questions you have generated.

The first step in creating a research plan is to identify milestones. Some milestones will be procedural milestones defined by your academic program. In addition, you will want to divide your research into milestones. One logical method is to do this chapter by chapter, in which case your milestones might be: Introduction/Problem Definition, Literature Review, Methodology, Analysis, and Conclusion. This is a good place to start, and might be suitable for the writing milestones, but is probably not sufficiently defined to truly get you through your research project. Most likely, the bulk of your research project is contained in executing your methodology; thus, you will need a set of milestones to get you through this portion of the research.

To identify your milestones, begin with your pseudo-investigative questions. Which of these are needed to provide you with background information? Which ones aid in designing the experiment? Most likely, answering these pseudo-investigative questions should be early milestones in your research plan.

Next, turn to your investigative questions. Are they all answered through a single experiment/data collection process, or do some require separate experiments/data collection? If more than one collection process will be required, your investigative questions themselves can serve as high-level milestones, with the lower-level milestones consisting of establishing the method, running the experiments, and analyzing the results.

If your investigative questions are relatively independent from each other, they can serve as separate sections in an analysis chapter, or separate chapters in a scholarly-format thesis or dissertation. Scholarly-format documents typically consist of introduction and conclusion chapters with the middle chapters consisting of separate publishable units, such as journal articles or conference papers. A well-crafted investigative question should be able to serve as an independent, investigatable unit, and thus could be its own journal article or conference paper. Investigative research questions that are the building blocks for other investigative research questions can often be appropriate for poster session presentations, conferences that require presentations only without proceedings, or conferences that accept papers based solely on an abstract review.

Investigative questions and pseudo-investigative questions can also be useful in identifying topics for a literature review. The pseudo-investigative questions can directly identify topics that need investigating, while the investigative questions indicate independent variables, dependent variables, and context, all of which could be appropriate literature review topics. In some cases, the methodology options indicated by the investigative questions may also be a relevant literature review topic.

The investigative questions can aid in building a chronology to your research, suggest literature review topics, and provide an outline and organization for the research document. These questions can provide a logical organization to carry the reader through your thesis or dissertation, and can aid in organizing your conclusion chapter. Revisiting each question and discussing the answer you discovered for each question provides a coherent summary for your research and enables you to emphasize your findings and contribution.

section three

*Planning: Making a schedule
and getting organized!*

chapter six

Scheduling

Just like in other parts of your life, you should have a goal in mind for your research project. It isn't just enough to say, "I want to defend by such and such date," or "I plan to graduate with my classmates." Although having the end in mind is a good start, you need to understand the other necessary parts of your project to attain your ultimate goal of graduation. Scheduling will help with that and this chapter will walk you through the process of developing an initial schedule and then keeping up with that schedule so that it is a useful tool for your research project.

Why schedule?

Scheduling is the process of allocating resources to an activity and outlining when this activity should be accomplished. It turns your research plan into an operational timetable. With a good schedule, you'll be able to know what tasks you need to accomplish, what resources it will take to accomplish these tasks, and when each task should be accomplished. When done correctly, you'll have different schedules of different scales (more on that later), a document that you reference and update on a regular basis, and you will know if you are ahead of schedule (or behind schedule).

This chapter will walk you through the basics of developing a schedule. Before putting pen to paper (or keystrokes to keyboard), there are some preliminary steps to take. We'll also discuss the two different ways of scheduling and discuss the benefits of considering different schedules for different time frames. Finally, we'll get into the gory details of the different kinds of schedules that you might want to choose for your research project.

But, before you begin …

Before you sit down in front of your computer to develop your magical document that will outline what exactly you need to accomplish, the resources you need, and when to accomplish it, there are just a few preliminaries you should take care of. We'll call this the "data collection phase" of scheduling. At the beginning of any project, you will always have limited information. This is no different in research, and the uncertainty might be even more magnified because you might feel you know nothing, or close to nothing, about conducting independent research, much less researching your specific topic.

To begin the data collection phase, have a discussion with your advisor focusing on expectations. What kind of work is expected of me? How often should we meet? What would you like to discuss when we meet? What are the work hours that you expect? What are the criteria that you will use to grade my work and evaluate me? Do you have an example schedule that I can build from? Having this discussion from the outset will do two things. First, it sets the stage of the advisor–student relationship. If you haven't chosen an advisor, this could help to inform you on whether or not you and the potential advisor are a "good fit." If you have already decided, this discussion will paint a clearer picture of what it will take to be a successful graduate student.

Second, and for the purposes of scheduling, you will be able to gather those short-term and long-term tasks which you will eventually build into your research timelines. Short-term tasks include preferred meeting times, what documents to submit and when, any classes that you should take, and any upcoming breaks when either of you will be unavailable. Long-term tasks include larger tasks that will typically require further breakdown into subtasks and are major objectives in the middle of the research program or might be accomplished as a culmination of the program. Publication submissions for journals and conferences, your first committee meeting, data collection, and your thesis or dissertation defense are all examples of long-term tasks. Collecting and understanding both kinds of tasks are necessary starting points in developing a research schedule.

Another source of information for developing your schedule is the basic documents you receive from your school about your graduate program. Program brochures, school catalogs, publicly available school calendars, and course schedules are sources of information that you can use to build your schedule. This may seem like an obvious step now, but collecting and organizing these papers and websites when you receive them may not be at the forefront of your mind as you juggle and sort through all the new information you are receiving as a new graduate student. It becomes very easy to focus on only class schedules and managing your time around these upcoming classes, causing you to lose focus on a major element of your graduate education: your research.

Finally, let us circle back to the advisor. One of the questions that you should be sure to ask during one of your initial meetings is "Do you have an example schedule that I can build from?" Asking for an example is neither trivial nor inappropriate. By asking for a schedule, and hopefully receiving one, you understand better what your advisor will be looking for in your own schedule. He or she may be accustomed to a particular format, a software package, or a specific level of detail. Meeting his or her expectations on this will only serve to enhance the communication between the two of you as you both work toward your own graduation. If the advisor provides you with a document that is, in your opinion, not very useful or does not have one to give, perfect! You can impress your advisor in short order by showing your planning and organizational skills through a well thought-out and developed schedule. Read on!

Two types of scheduling: Forward or backward?

By now you might have a pile of notes of dates to complete tasks or general ideas swimming around your head of when to complete what and might be thinking: "Great! Let's build a schedule around these dates." Although a completely valid approach, pause and consider what your goals are for the research and education. Understanding what your end goals are for your education will guide you to using either the backward scheduling technique or the forward scheduling technique.

The forward scheduling technique is the more intuitive approach in scheduling and so let's discuss that first. To forward schedule, you simply have to work forward from the current date. As you have a collection of tasks and a general idea (or at least a good guess) of how long these tasks will take, you develop your research schedule by placing one task after another. As the name suggests, you start from the beginning and move forward from there. The completion date is determined when the last task in the schedule is complete.

In forward scheduling, the key pieces of information that you need are the durations for every task in your schedule. As there is a great amount of uncertainty in how long tasks will take, this technique may produce surprises in how long the research and education may actually last. You may surprise yourself when you complete your initial schedule or may be shocked to see that your graduation date regularly shifts to the right. This is not to say that forward scheduling is a poor method of scheduling and should be avoided. Rather, the circumstance of your educational goals should be evaluated. Are you a working student who can afford the additional time in your program? Do you have an "open" schedule with no hard deadlines to graduate? What responsibilities outside of school do you have that limit the amount of time in any one semester or quarter for you to work on your research and studies? These are examples of factors to consider at the outset of creating your schedule. If you find that an open-ended schedule for graduation makes sense, then the forward scheduling technique is appropriate for you.

For full-time students and those with "hard" deadlines for graduation, backward scheduling may be more appropriate. In this less intuitive method, you plan with the goal in mind and work backwards to the start. That is, you set your graduation target date and create your schedule beginning at the end. Task duration remains a key piece of information, but "force fitting" durations for tasks will occur in backward scheduling. That is, you bound or expand the amount of time spent on each task in order to make your schedule work. Understanding and maintaining the logical sequence of tasks remains important, after all you can't collect data on your project if you haven't developed the research question. This potential to force fit tasks into time will require you to be more disciplined in keeping with your schedule. Backward scheduling might be advantageous in dealing with the amount of uncertainty in the task durations. Since you are never 100% certain how long a task duration will actually be, this method of scheduling not only allows you to be more flexible in the amount of time you allocate per task, but with a goal in mind you must also be more disciplined in completing these scheduled tasks.

Consider scale

One of the difficulties that you might encounter in developing a schedule is the overwhelming amount of detail that you might feel necessary for a good schedule which at the same time makes it unwieldy. The trick to scheduling is having a document or, as we'll make the case in this section, a series of documents that give you just enough information to track the completion of what you need to do. Having a single "master schedule" is a common method where you have a single document with every single task. Unless you have a plotter to print on poster-size paper, you will never be able to see the entire schedule and the final product will just confirm your intimidation with the whole research process.

Instead, consider scaling your scheduling efforts between a high-level, low-detail "master schedule" and lower-level, high-detail "short-term" schedule. Just like in mapping, the different scales will let you see different levels of detail as needed. This approach has two advantages. First, you'll be able to see the bigger picture much clearer. Having a document that you can easily digest visually might put your mind at ease while at the same time you can see those large challenges ahead of you. Second, you can go into as much or as little detail as you need. You might have friends who feel like they need to plan every small detail and their detailed plans give them the sense of control they need to move ahead. For others, too much detail might make their head swim, and so they might create schedules with just enough tasks to break down the big challenges and move on with the research project.

For the schedules, we recommend this two-phase approach: the big picture and then the details. For the big picture, your master schedule, you should develop something that can fit onto a single sheet (or maybe two-taped sheets) that you can view in a single glance. This document should capture all of those major milestones, big rocks, and significant tasks. You not only should be able to view this with a single look, but you should also keep it in a place where you can see it every day as a positive reminder of the progress you are making.

For the detailed schedule, you can approach this in one of two ways. If you are task-oriented, take those major milestones and break them down into further subtasks. This is the work breakdown structure (WBS) method which we will get to shortly. A second way is to plan around your school schedule. Your program will naturally be divided into quarters or semesters which are neat little blocks of organization around which to plan. With your master schedule, identify which tasks you need to complete in the upcoming term and develop your more detailed schedule for the relevant portion of the master schedule. By approaching your schedule from different levels of scale, you'll be able to see the big picture and at the same time have a schedule that outlines the details at an actionable level that is right for you.

Milestone schedules and the work breakdown structure

Milestone scheduling

Having collected as much data as you can, and having thought about the appropriate schedule type and levels of detail, you are ready to develop your first schedule: the milestone schedule. This step is a very simple way to organize all those thoughts, calendars, and notes you have collected until now. The milestone schedule is a very-high-level view of the overall project ahead of you. It contains the basic points of your program: project start, project end, major events along the way, and any important deliverables you have identified.

Major events and important deliverables are, admittedly, very vague things for you to consider. But at the start of every project, you have limited information, so follow this rule: If it sounds important, it most likely is. It won't hurt or waste any additional time to include these uncertain tasks. A task's importance will reveal itself as you develop your schedules further and you can always refine the list during the course of the project.

One step to help in determining importance is to consider tasks and timelines that are outside of your control. For example, the process for submitting and approving your final thesis or dissertation document has procedures and deadlines you will be required to adhere to. The process for applying for graduation is another. Is there lab and safety training that you need to accomplish before accessing certain facilities? Or how about specific courses that you need to complete in support of your topic? Other milestones might include

- Identifying your research question
- Identifying your research advisor
- Holding your first committee meeting
- Taking and passing your qualifying exam
- Submitting various drafts of your thesis or dissertation
- Submitting your research plan for Institutional Review Board (IRB) approval
- Defending your thesis
- Applying for graduation
- Graduation

A simple table or bulleted list with completion dates will suffice as in Table 6.1. Creating the milestone schedule is a process in which you have identified the critical tasks ahead of you, found their due dates, and organized them in such a way that you have these critical tasks in a single, easy-to-read-and-access document. The milestone schedule will be the basis for other working schedules you create, so a little time spent working on this document will help immensely as you move to the next step in scheduling.

Work breakdown structure

The WBS is the next step in your project planning. The WBS is a work structure that accounts for more detailed elements. The goal of this step is to break down the high-level, broad activities developed in the milestone schedule into a comprehensive structure of small elements of work, hence, the name of this step: work breakdown structure. The WBS is more than just a list of small elements: the arrangement of these small elements will show how each activity relates to other activities within the structure. Research projects are well suited for the implementation of a WBS.

By creating a WBS, you increase the likelihood of accounting for all of your research projects activities, both big and small. At the early planning stages, developing the WBS may seem like a daunting and minutia-filled task. So we refer back to the issue of scale. Recall that we suggested viewing schedules according to different timescales. When timescales are short—like for a given academic term—a more detailed view is appropriate. For longer timescales, like your entire research plan, taking a high-level, less-detailed approach will be just fine.

Table 6.1 Example of level 1 WBS

Level 1	Description	Start date	End date
	1 Classes	August 2016	May 2018
	2 Research	September 2016	May 2018

The WBS method is helpful in this regard because it allows you to break down a project according to different levels of scale that will correspond to different timelines. Individual components in a WBS are referred to as WBS elements, and the hierarchy of each is designated by a level identifier. Elements at the same level of subdivision are said to be of the same WBS level. Descending levels in the WBS provide increasingly detailed definition of project tasks. The complexity of a project and the degree of detail that you want will determine the number of levels in your WBS. Each task is successively broken down into smaller details at lower levels of the hierarchy. You would continue the process until the hierarchy reaches specific project activities (that is, WBS elements or tasks). The structure of the WBS looks very much like an outline. But we should be emphasized that the WBS is not a simple outline, it is a very specific breakdown of the work required for your project. The basic approach for preparing a WBS is as follows:

- *Level 1 WBS.* This contains the final goals and the largest elements of the project. This item will not have much detail. Consider level 1 WBS elements as "big picture" items.
- *Level 2 WBS.* This level contains the major subsections of the project. For your graduate education, you might want to refer to department handbooks, program guides, or even the school's catalog to determine the major milestones you need to consider.
- *Level 3 WBS.* Level 3 of the WBS structure contains more definable components of the level 2 subsections. At this level, you should begin to have a sense for the resources needed (including time) to accomplish the specific element.

Subsequent levels of the WBS are constructed should you need more specific details. Again, the number of levels and amount of detail are dependent on you and your desire to understand and control your project and timeline.

As an example, consider Table 6.1. For this project, we've determined that the two most significant ("big picture") project elements are classes and research; we use these elements as labels for the first level tasks. Note the numbering, the description, and the end dates for either task. For a WBS, each element has a distinct number, a clear and concise description, and well-defined start and end dates.

To develop the second level of the WBS, it is a matter of determining the logical breakdown of each task. For the current example, the "classes" task is broken into courses to take according to the quarter and the "research" task is broken into the basic elements of the research process. Table 6.2 provides an example for the second level.

Table 6.2 Example of level 2 WBS

Level 2	Description	Start date	End date
	1.1 First quarter classes	August 2016	October 2016
	1.2 Second quarter classes	October 2016	December 2016
	1.3 Third quarter classes	January 2017	March 2017
	...	...	...
	2.1 Introduction	August 2016	October 2016
	2.2 Literature review	August 2016	December 2016
	2.3 Methodology	January 2017	March 2017
	2.4 Analysis	May 2017	August 2017
	...	...	...

For each subsequent level, you would continue the breakdown of each element until you get to a point where you can begin to measure completion of the activity. In project management practice, a WBS might go up to six levels deep wherein tasks might be broken down into further levels of detail. The impetus for doing so is primarily the large amount of work and tasks that need to be accomplished and coordinated across the project. For your particular project, coordination across project teams should not be necessary. Classwork and your research are, for the most part, an individual effort. Therefore, six levels may not be necessary, but stopping at too few levels will not provide the right level of detail to track progress. Appendix 6 gives you a sample WBS to get you started.

So what then is the right amount of detail? Borrowing from another area of management, the concept of S.M.A.R.T. goals provide guidelines in knowing that you have the right detail in your WBS plan (Figure 6.1):

- *Specific*: Is the individual task written so you know the exact outcome?
- *Measurable*: What is the item that you can count which tells you progress?
- *Attainable*: Do you have the necessary resources and ability to accomplish the task?
- *Relevant*: Is the task properly related to its parent task and the overall project?
- *Timely*: When will this task be complete?

Following the basic guidelines of S.M.A.R.T. allows you to treat each WBS task as an individual goal to accomplish. "Specific" and "measurable" will tell you what exactly needs to be done and if you are actually making progress. "Attainable" addresses the issue of resources and your own abilities to accomplish a task. "Relevant" ensures that each task is pertinent to the overall project and that you are not wasting resources.

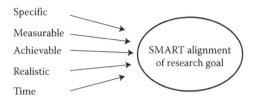

Figure 6.1 SMART alignment of research goal.

Finally, "timely" sets clear deadlines for when tasks should be accomplished. This prevents a particular task from lingering and or procrastination behaviors to take over. Thinking through and applying each element of S.M.A.R.T. will help in developing a well-structured and appropriately detailed WBS.

In our advising of students, we've found that three to four levels in a WBS are adequate to meet the basic elements of S.M.A.R.T. and to get started for research and classes. Going beyond four levels might be necessary when completing smaller-scale, shorter-timeline schedules. To get you started, we've provided a sample WBS for the thesis portion of the work ahead of you (Appendix 10). Note that the sample stops at four levels and that further development of the WBS may be necessary to outline each task.

Graphically charting your schedule

Understanding your program's milestones, breaking down the work into smaller and measurable elements, and assigning start and end dates to these elements now give you a more detailed schedule. Up to this point, your schedule has been a list: A list of tasks and a list of dates. Stopping here may be okay for you to use, but there are two more tools that we would like to introduce for use in scheduling in your program. The Gantt chart and network diagramming methods are two ways to graphically depict your program and provide the monitoring and control needed in project management.

Gantt chart

The oldest and most widely used method for depicting a project schedule is the Gantt chart. The Gantt chart is sometimes referred to as a bar chart or timeline. It shows when each task will begin, how long it will take to complete, and you may be able to discern dependency relationships from the tasks. It is very simple to create and easy to understand which explains its wide utility in project management.

Figure 6.2 provides an example Gantt chart with its basic elements. It shows the title of activities on the left; time is indicated at the bottom; the planned execution of each activity is indicated by the location and size of each bar; and the current time is shown via the vertical line. Some specialized scheduling software may show dependencies (arrows among bars) and distinguish between tasks (horizontal bars) and milestones (diamonds or some other symbol).

Advantages of employing the Gantt chart include ease in creation and understanding, ease in portraying progress in the schedule, and ease in modifying tasks to include or remove detail. There are many software products that can produce Gantt charts quickly and effectively, but you do not need to purchase specialized software for this kind of charting. Common spreadsheet applications and word processing documents can easily be used to visually depict a schedule.

The Gantt chart method is not without its limitations, however. The primary limitation is the chart's ineffectiveness to show activity dependencies. Without these dependencies, the results of schedule changes from delays or early completions are not easily shown. Activity paths, the chain of events for a particular activity, are not easily discernible and so impacts from delays in one part of the schedule are difficult to understand. Understanding these impacts is particularly important for an important activity path, the critical path.

The critical path

A schedule's critical path is the chain of activities that dictate the earliest possible finish for the project. It is the path through the project schedule in which a delay in any given activity will result in a delay for the overall project. In project management parlance, the activities along this path have zero "slack" or "float." Other activities in the network that are noncritical will have some "slack" or "float" meaning that these can be delayed without affecting the overall completion of the project.

Within the research setting, it is important to know the critical path so that you can forecast the tools, methods, and classes that you might need before getting to some activity. For example, if your research calls for understanding a particular simulation method, it will be good to understand at what point in your program this simulation method will be covered and to plan for completing your particular simulation after it is covered in class. If you find that the class offering is too late in your program, you may have to make plans to independently study the method on your own or choose a different method for your research.

So how do you determine the critical path? You must know the activity dependencies. Recall your milestones and work breakdown structure. Using this list, assigning which tasks precede and follow each other will provide part of the data necessary to determine the critical path. The other

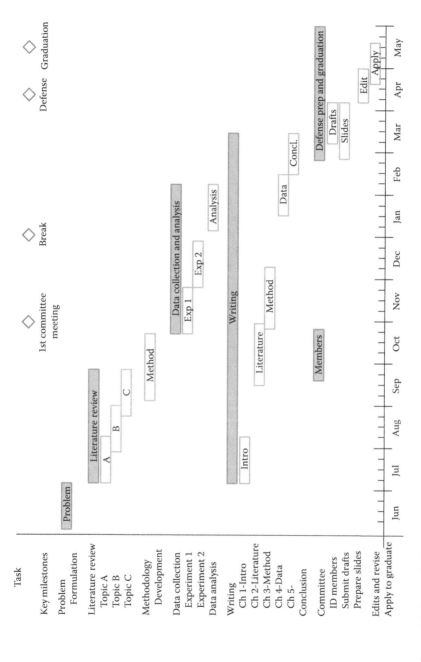

Figure 6.2 Example Gantt chart.

Table 6.3 Example precedence table

Label	Task	Estimated time	Preceding task
a	Problem formulation	4 weeks	–
b	Literature review: Topic A	4.5 weeks	a
c	Literature review: Topic B	4.5 weeks	a
...	...	...	...
i	Writing: Introduction	4.5 weeks	a
j	Writing: Literature review	6 weeks	c, i
...	...	...	...
s	Post-defense: Edits	3.5 weeks	r
t	Post-defense: Apply for graduation	4 weeks	s
u	Graduation	1 day	t

data necessary are estimates for the durations of each activity. This time span is necessary for calculating the length of time for each activity path and then determining which of these paths is the critical path. Table 6.3 provides an example to organize these data points.

Network diagramming

To determine the critical path for your project, another widely accepted graphic must be used: the network diagramming method. There are a number of different methods within network diagramming which include:

- Program evaluation and review technique (PERT)
- Critical path method (CPM)
- Precedence diagram method (PDM)

Each method outlined earlier has nuanced ways for depicting the activity network for your project, but in general they share many basic features. First, unlike the Gantt chart, activity dependencies must be known. Without knowing these dependencies, making the connections between tasks to form the network is not possible. These dependencies form a chain of events and will allow you to trace individual activity paths and downstream impacts from delays or early finishes. Second, the duration of each activity is known so that the time span for an individual path can be calculated. Together, knowing activity dependencies and durations will allow you to see the results of a change in one part of the schedule on activities downstream in the network.

Determining the impact of a change is accomplished through network analysis techniques. Not only can you see the impacts of changes in start and end dates as well as changes in overall completion, but network diagramming also makes it possible to calculate different metrics in your project such as schedule performance, risk, and even conduct simulations or "what if" scenarios. Many of these features of a network diagram schedule are advanced techniques and although we'll leave it to you to review the details in other texts, we've provided a quick overview below. In short, a network diagram will help you

- See activities that immediately precede a given activity
- See activities that immediately follow a given activity
- See activity paths (a string of activities) that run parallel to each other
- And, most important, find the critical path (the path that, if delayed, delays the entire project)

In order to see these things, an activity dependency list as part of a precedence table (Table 6.3) is needed to develop the network diagram in Figure 6.3. Visually, you might decide that it is no different from the Gantt chart. After all, in the Gantt chart, you can still discern some parallel activities as well as clearly see a logical sequence of events for your research. However, because it does not incorporate dependency activities, the Gantt chart cannot show you the impact of a change in your schedule on the overall project and the different paths (both parallel and the critical path) might be impossible to determine.

The activity dependency list in Table 6.3 provides the logical connections for each activity in the network. Also, the durations are known (or you at least have good estimates) and with a start date for the project, you can overlay the activities and their connections onto a timeline (the bottom axis of the chart). For our network, you can see that there are three distinct activity paths. Summing the total time along these three paths provides three time estimates; the path with the longest time span is your critical path. We've highlighted that path in Figure 6.4.

Where the parallel paths merge with the critical path are areas where "float" or "slack" time can be injected. In the critical path figure, activity "Writing: Method" is a merge point where one of the parallel paths joins into the critical path. The critical path takes 30 weeks to get to that activity, while the activities on the parallel path take a total of 21 weeks. Taking the difference between the two shows 9 weeks of excess time. This means that the activities on the parallel path (up to the merge point of "Writing: Method") can be delayed by up to 9 weeks without affecting the schedule. The excess time is known as "float" or "slack" time. The critical path does not have any float or slack time and so any delay on the critical path will cause a delay in the overall project, hence its label as the critical path.

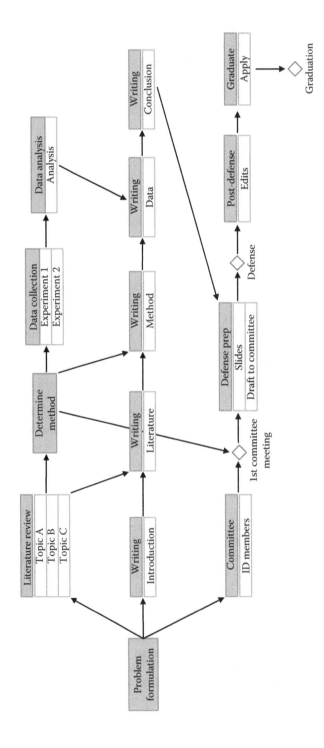

Figure 6.3 Example network diagram.

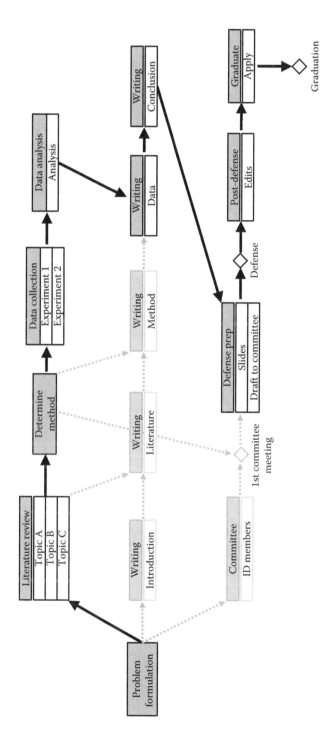

Figure 6.4 Example network diagram with critical path.

A final note on the critical path: It moves! If unexpected things happen in your research program and an activity consumes all of its float time, then the critical path may shift to what was previously a parallel path. Having shifts in your critical path might be an indicator that you do not have as good a handle on your schedule as you thought, and a delay in your program might be inevitable. The point of producing a schedule in the first place is to plan out your work, monitor it, and provide you the ability to control your schedule. When your critical path moves, all is not lost—it's an indicator of a potential problem. With a good schedule, not only do you have a plan to reach your long-term goal of successful graduation, but you can also be in a position to identify potential problems early and address them in your program.

chapter seven

Organization

By now, you have decided on attending a graduate school, enrolled, and maybe even have completed your first year of classes. Your schedule for graduation seems to be on track and maybe you have a few ideas for your thesis research. As a student, you wouldn't have made it this far without some organizational skills mastered, and definitely as a professional with a full-time job, you have some tried and true techniques for organizing your work life. But graduate studies and research, if you are like most people, should be unlike anything you've experienced before, and a question that you might have is how to manage all of the things that go into successful research. This chapter will help with organizing your graduate research.

Organize what and organize how?

One of the very first things that you should know about organization is that different techniques work for different people. It's ironic that I (Vhance Valencia) should be writing this chapter on organization once you view my workspace, desktop space, and the spaces that are designated as "mine" when you come into my home. My wife and the coworkers around me are far more organized, but their spaces are a reflection of how they work best.

Take my spouse, for example. One of the things that I love about her is her need for neat and tidy spaces. Certain spaces in our home are well arranged: you know exactly where to look for certain items, quick glances tell you what things are running low or missing, and you simply get a feeling of orderliness and tidiness. Now, I'm not sure what exactly motivates her to have these extraordinary organizational skills. Maybe she finds an inherent value in having a tidy home or maybe she's trying to avoid the stress of having a disorderly home—maybe we can all vouch for the increased stress when there is a disorderly home but whatever the motivation, her techniques for organizing the house are ones that she has come to on her own. Just like my spouse's skills are fit for her, you should work to develop organizational research skills which will be fit for you. When it comes to organization, different methods work for different people. So organize your work the way it suits you personally best.

Organizing the things that go into your research will be very different than organizing your notes and work for your classes. In fact, organizing for classes can be comparatively easy; after all, a teacher worth his or her salt will provide the class with a syllabus outlining what to expect, when to expect it, and different modules that will make up the class. From here, you as a student can organize your physical and digital files to match the class requirements.

But in research, you won't have that syllabus document to organize your files around. You won't have neat little blocks of learning and chunks of time that can easily be the labels for your class notebook or series of files in a hard drive. What you will have are collections of papers, ideas, and lots of analysis that will end up contributing in some way to the final document that you are to produce. Finding a way to organize these things will help make the research process (and writing process) be as efficient and smooth as possible.

The research itself

Keeping the end in mind, let's consider the organization of the final thesis document. After all, all of the research work that you will embark on culminates in this single document. There are two strategies that you can use: the traditional five-chapter format or the newer scholarly article format which is gaining popularity. The decision on which format to use will depend on two factors of your end goal with the research and your thesis advisor's preference. This section will cover both formats within the context of these two factors.

The traditional five-chapter thesis is the method that the coauthors were taught as graduate students working through their own research problem. The five chapters document five distinct stages in the research process. These five chapters are:

- Chapter 1: Introduction
- Chapter 2: Literature Review
- Chapter 3: Method
- Chapter 4: Data and Analysis
- Chapter 5: Discussion and Conclusion

Considering the research process, you'll see how these chapters fit together and the role that they play in answering your research question. In Chapter 1, Introduction, you are exposing the reader to the general context of the research and ultimately your research question and supporting questions. Chapter 2, Literature Review, explores the different areas of the topic that you are working on and examines what research has been conducted in the past. Chapter 3, Method, outlines the research method that your work employs. Will the work be quantitative or qualitative? What is the research design that you will employ? What are your data sources? Who are the participants in the study? Questions concerning how you will go about conducting the research are answered here. Chapter 4, Data and Analysis, describes the data that you were able to collect and carries out the analysis that you described in the previous chapter. Finally, Chapter 5, Discussion and Conclusion, puts all of the previous four chapters' work in context relative to the original research question, summarizes the results, suggests future work, and describes the applicability of the results to the more general context of the problem as it was articulated back in Chapter 1.

As an experienced researcher, the five-chapter format feels like a very structured approach, and this structure is useful in articulating the end state for the students. If you picture two funnels joined together at the narrow ends, this format takes a big view of the problem, narrows the scope to the specific research question, and then applies the findings back to the big view of the problem (Figure 7.1).

A second method to organize your thesis document is the newer scholarly article thesis. In this structure, the focus shifts from an organizational method that steps the reader through the research process, but rather on a document or two that will be used in a publishing venue. Consider the scholarly article thesis as a "document within a document."

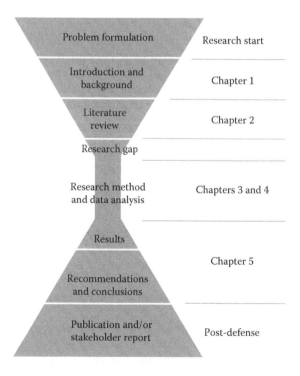

Figure 7.1 Two-funnel model of research.

Depending on how prolific a writer you are, your thesis might contain one, two, or three documents that have high potential for publication. In this format, you will not have a set number of chapters, but you do have a general format to follow:

- Chapter 1: Introduction
- Chapter 2: Literature Review (optional)
- Chapter 3: Article #1
- Chapter 4: Article #2
- Chapter 5: Summary
- Appendices: Extended Literature Review, Extended Method, etc. (optional)

Although the general research process of introduction, literature review, methodology, analysis, and conclusion are not explicitly followed in this format, the research steps are contained in each of the articles produced, albeit in condensed form. As well, you may find redundancy in literature reviews and so placing the full literature review as an appendix is an option. This format's focus is on the articles within the thesis and dissertation.

So why the different formats? These formats exist because they approach the goal of academic publishing differently. Recall that choosing between one of these two formats is dependent on the publication goals and preferences of the student and advisor. The traditional format takes some additional work to produce a publishable article, but it is very good for stepping through one's research program. The scholarly article format is less structured, but results in one, two, or three documents that have good potential for publication. The scholarly format is particularly well suited to research questions composed of separable, semi-independent investigative questions. The traditional format, on the other hand, is well suited for a single, big problem that does not have clear divisions into smaller pieces. Although most research advisors have the goal of publishing your research work, after all our tenure is based on our ability to publish, sometimes the unstructured nature of the scholarly article format might not fit the student or problem at hand, thus calling for the traditional format. Or, you as a student might have a natural penchant for research work and might be able to pursue several research questions and the scholarly article format might be the best method for capturing your work.

The organization of your thesis will be based on your own as well as your advisor's preference. As with other elements of your graduate education, you should sit and discuss the final form of your thesis early so that you have the end in mind.

Literature search articles

One of the major elements of your work will be the literature review or literature search. In this step of your research, you are trying to understand what research work has already been accomplished with regard to your topic. You'll read many, many publications, amass a large collection of these publications (either paper or digitally), and so long as you understand what you are reading, many different thoughts and ideas about how each piece relates to each other and the specific question you are trying to answer. The goal with the literature review is that you develop an in-depth knowledge of the field that you are studying. So how do you organize this all?

In the literature search, there are really two things that you need to keep organized: (1) the ideas and key points in each article and (2) the citations for each piece of work. Both are critically important to stay on top of.

In organizing ideas, one common strategy is to use a *matrix of key points* that each article presents. The purpose of the matrix is to have a space where you can visualize how each reading relates to some key theme in your research topic. One set of headers in your matrix would be the individual themes you've identified, while the second set of headers identifies the article that you've read. Where the topic theme and article identifier intersect is a place for you to write how the article addresses that theme. Figure 7.2 is an example from the work that coauthor Vhance Valencia did from his own master's research.

As you can see, each cell indicates how the authors fit into the topic theme. Some authors did not address the topic and so it's just fine to leave the cell blank. The matrix method is useful because key points on each theme are easily stored. Visually, you can see the extent to which an author addresses different areas of a topic or how much of a theme is addressed by many different authors. Creating the themes in your matrix, in and of itself, is a useful exercise as you now have more purpose in your reading and can extract only what you need from each article. Also, consider the matrix as flexible: you can add and remove authors and themes as your literature review and understanding of your topic increases and is refined.

Another common strategy in organizing the ideas in a literature review is the *notetaker method*. This method is a single page (or even less!) summary of what you just read. The notetaker should contain basic bibliographic information (title, author, publication, date, place of publication, etc.) and a summary of the key points of the article. The bibliographic information is for your use later on when you are documenting your citations and for easy retrieval of the original work. The summary of the key points is for your use in relating the work to your research effort. Although the topic may seem relatively easy to remember, with all the reading that you'll do, you'll quickly forget having read the article to begin with.

	Study method	Problem solving	Admin. ability	Supervision and team management	Interpersonal relationships	Other	Knowledge	Experience	External factors
Bowenkamp and Kleiner (1987)	Opinion based	✗	✗	✗	✗	✗			
Pettersen (1991)	Meta-analysis	✗	✗	✗	✗	✗			
Anderson and Tucker (1994)	Survey		✗	✗	✗		✗	✗	✗
...									
Crawford (2000)	Meta-analysis	✗	✗	✗	✗		✗		✗
Hauschildt et al. (2000)	Survey	✗	✗	✗	✗	✗	✗		
Odusami (2002)	Opinion based, ranked	✗	✗	✗	✗		✗		
Hyvari (2006)	Survey			✗			✗		✗

Figure 7.2 Example literature review matrix.

Also, avoid the temptation to copy just the abstract of the work you just read. Although the abstract does serve as a summary of the article, it is imperative that you go through the work of making connections between what you have read and your thesis topic. An example notetaker and the headings that you might use is given in Figure 7.3.

Beyond organizing the ideas that you are discovering in your literature search, you'll find that there are other things that you should also keep organized. Search databases, keyword lists, and your electronic files systems are other items related to conducting your search:

- Know what databases are available to you. Google and Google Scholar are great tools to use, but they should not be the sole source for conducting research. Internet search engines are a good starting point, but library databases will generally have more complete access to articles and will return different search results.
- A keyword list is your repository for those key terms that are relevant to your topic. Knowing the right keywords will result in relevant search results, but how do you know which keywords to use? After all, if you don't know the right keywords, how will you get the relevant publications? Like the theme matrix, your keyword list will shrink and grow as you conduct your literature review. As you come across relevant papers in your literature review search, be sure to examine the key terms tagged for those papers—you may discover important keywords that you hadn't originally thought to include.

```
                                          Date read:
        Title:
        Authors:
        Category:
        Complete citation:

        Description:


        Key findings:
            •
            •
        My thoughts on how this applies to my research:
            •

            •

            •

        File saved at:
```

Figure 7.3 Example notetaker.

> Eventually, you'll learn what the good keywords are and which are irrelevant; keeping a keyword list will help in this process and mini-mize the number of repeated searches that you do. Be sure to include which databases you have searched in for the particular key terms, in order to avoid misses and duplications.

- Having a good electronic filing system for your articles will be important as you amass the different publications you have read or are planning to read. First, note that we are advocating using *electronic* files and not paper files. Electronic files are much easier when it comes to organizing. They reside digitally and are therefore more compact, you can easily search the contents of all of your files for a particular keyword or phrase, and you can replicate or delete copies to fit whatever organizational structure you choose. Just be sure that your folder structure makes sense to you and the filenames that you choose are meaningful.

Citations: Organizing information about information

Tied to your literature search is the organization of the citations for those articles that you are collecting. Organizing these citations is a critical step in staying organized in your research. You may often feel the need to just "plow through" your reading list so that you can gain as much knowledge on the literature in as short a time as possible. And, in the short term, this tactic works: you'll amass a great amount of readings completed and will have grown your knowledge and expertise in your topic exponentially. But in the long term, you'll experience difficulty. You'll find yourself spending hours researching articles solely for their bibliographic data because it is simply impossible to remember all the details needed in a proper citation. And once you do find the correct articles, the process of collecting citation after citation is quite frankly, tedious. Finally, composing the final bibliography to include at the tail end of your thesis adds more tedium and time if you choose to create this part of your thesis manually.

So what should you do? One of the most important tools that you will use in your graduate research is a good citation manager. A citation manager is a tool that allows you to save and organize article citations and create bibliographies and in-text citations as you write your document. All university libraries will have support for at least one citation manager platform, but you can also purchase your own software or subscription service. So the place to start is at your school library and sign up for an orientation for the available citation managers.

Although all citation managers have the same basic features of saving and organizing citations and creating bibliographies and in-text citations on the fly, not all managers are created equal. There are other features that you should consider when selecting the right tool. Think about how you plan on working: will you be working from multiple computers or your own? Will you be working with a group of people? What type of word processing platform do you prefer? Also consider performance features: what kind of Internet connection does it require? What is the cost for the software? Is there a subscription fee? Can it import from databases and webpages? What kind of storage capacity is available to you? Can you store articles, notes, and/or tags directly in the citation manager?

These features are hard to assess if you are researching this tool on your own. Good sources to determine what citation tool to use are the librarians at your school and fellow classmates. Use the resources around you! Below is a quick list of citation managers that might get you started:

- citeUlike (free; web-based)
- EndNote (commercial; Windows, Mac)
- JabRef (free; Windows, Mac, Linux)
- Papers (commercial; Windows, Mac)
- Mendeley (free; Windows, Mac, Linux)
- RefWorks (commercial; web-based)
- Zotero (free; web-based)

Organizing your spaces: Virtual space and physical space

In this final section of the chapter, we get to organizing the two different spaces in which you'll work: virtual and physical spaces. Virtual space refers to areas in your research in which you'll store, retrieve, and work on the things that will progress your research. It's important to think about a virtual space as your work area because you will no doubt be using a computer with some access to the Internet. If you do not properly manage your virtual space, it can be as much of a hindrance to your work as an unorganized file cabinet, incomplete notes in a notebook, or just a messy desk. The point in organizing these two spaces is to minimize your distractions, which will increase the likelihood that you'll be a productive researcher and student.

Physical space

As a graduate student, it will be very important that you have a regular place that you'll be able to call your own. As an undergraduate student, finding a clean desk to work at probably sufficed because you were able to carry all of the materials that you needed to study with. As a graduate student conducting research, it's unlikely that you are as mobile with your work. You might be tied to a lab for some experiment, have data in the form of survey results, or simply have too many references to be carrying around. So, as a graduate researcher, you will likely find a need to have a single space from where you can work.

Whether you choose to set up a home office or are provided with an office space on campus, this space should be organized so that you can find all your papers, books, journals, and draft work easily. If a filing cabinet is available to you, hanging file folders and a label maker to create easy to read tabs will help in organizing the papers that you'll collect in the next months. What to call your tabs? Refer back to your literature review matrix for the common themes and use the different chapters from your thesis as a starting point. As your research goes on, you'll get a better feel for what these different categories should be. Setting up a system for your physical space is important because you will minimize the amount of time you'll spend looking for that last journal article or draft paper that you wrote. Less time spent looking for lost items will equate to less stress during your research work.

Once you have your system set up, you have to do regular maintenance on your physical space. Just a small amount of time (as little as 10 min) each day will go a long way in eliminating distractions and reducing frustration. For those that find this task too burdensome, consider setting a timer to limit the amount of time spent on this task. Tidying up and filing away your collections of papers will help in keeping your momentum with your work.

Virtual space

Much like your physical space, organizing your virtual space is also intended to reduce the distractions in your work. When deep concentration is necessary, actions like shutting down your email program and getting off social media will go a long way in making you a more productive researcher. Messaging applications like these serve as interruptions and make you a less productive individual. One way to manage your virtual space is to employ a two-browser strategy.

Think of the two-browser strategy as using one type of browser for work and a different browser for play. In today's Internet browsers, you can store bookmarks for websites, RSS feeds for new information, and store login information for social media sites. Making it more difficult to access personal items like an online shopping account and your social media and news accounts decreases the likelihood that you'll click on these links and possibly waste a half hour or more of your work time. Using one Internet browser for your research work and another browser for personal or "play" time, will help keep you on task and, at the same time, allow for the conveniences built into modern Internet browsers.

And just like your physical space, your virtual space also requires maintenance time. One very important key is to make sure that your files are backed up regularly. There is no worse feeling than realizing you've lost weeks' or months' worth of work because your hard drive was stolen, dropped, or crashed. The best method is to have an automatic backup run, maybe through a cloud storage system. Computer operating systems today have built-in backup capabilities, so be sure to take the time to learn about and activate this feature.

Another important key in maintenance is managing a folder system that you can navigate easily. In your research, you'll not only amass digital publications for your literature review, but you will also need a place where you can store your written work, data, analysis of that data, computer code, and all other digital files that result from living and working in a computer connected world. Try to keep your folder structure organized logically so that when you are looking for something in particular, it is obvious where you need to click next. Your logic will be your own and so the organization of your files will be your own as well. A good starting point would be to review your work break down structure and go from there (see Appendix 6 for an example). And, whatever you do, avoid creating a "MISC" or "Miscellaneous" file! These file folders end up being just a collection of things that you don't know what to do with.

Finally, email. As a student, your email inbox may start out to be very manageable. But, as you move along in your program, and if you ignore email maintenance, you may find that your inbox becomes full, may get locked because of its size, and in general unmanageable. If this describes you, don't worry; problems with email aren't confined to your inbox.

Everyone has issues in dealing with their email system and we propose a few strategies for your email:

- Just like the two-browser strategy, have multiple accounts designated for different areas of your life: school, work, personal, and junk. After setting up a junk email, you may soon find that this is the most active of your email accounts.
- When you receive an email, immediately do one of the four things: act on it, forward it, archive it, or delete it. Taking one of these four actions will help in eliminating a backlog of emails in your inbox.
- Strive to keep no more than 10 emails in your inbox. If you get lots of emails in a day, look to see if there are emails that belong in groups (i.e., project related, class related, from specific individuals) and place like emails in specific folders to deal with all at once.
- Set up rules to automatically deal with common emails that you receive. For example, your social media accounts should be automatically deleted; any notifications belong in the application itself, that's why there's an app to begin with! If you have your group folders set up (see previous tip), set up a rule to automatically move emails from your inbox to that folder.
- Finally, don't spend all day in your email inbox. Block off specific times of the day to reading/responding to/filing emails. More on this in the time management chapter. Managing email does not equate to productivity; so spend your time on energy toward your research and writing to move you on the path toward graduation.

In this chapter, we walked you through some strategies for organizing your research work. If you are in the middle of your program and feel like your work is a disorganized mess, picking just one area that we discussed and spending as little as 10 minutes a day toward organizing will set you on a path to getting your work under control. You'll see that having an organized system will make you feel that your research life is in order, under your control, and that you are on your way toward graduation.

section four

Project execution and control

chapter eight

Time management

You've preplanned, planned, and organized, and are now in the middle of your program. In project management, execution and control refers to the stage in the project where the work happens (execution) and making sure that the work remains within scope, on schedule, and within cost (control). In this stage, project managers monitor their project progress and make necessary changes in the project schedule and resource allocation. Treating your graduate education just like a project will help you stay within the bounds of your research project (stay within scope), graduate on time (stay on schedule), and not pay for any additional credit hours because of a delayed graduation (stay within cost).

Of course, you'll have to manage many different elements of your graduate education in order to graduate on time: pass and excel in your classes, find a thesis topic and advisor, and put in the energy and effort into your writing and research. Underlying each of these elements is your ability to manage your time wisely. Combining good time management skills with a well-developed schedule (see Chapter 6) will help you toward success in your graduate research program. Time is a precious resource; after all, whatever time you use, you'll never get back.

How is time management different for a graduate student?

Time management in graduate school will be quite different than that as an undergraduate student. For most people, you'll be tackling your graduate program a little later in life and quite possibly have a full-time job. Some of you might be married and have young children at home. With work and your personal life, your research work and studies will not be the only things to focus on. Each of the roles that you hold, whether student, worker, parent, or spouse, takes precious time. With some years and experience under your belt, you might now be thinking to yourself: "I have a lot going on. Where am I going to find all the hours in the day?"

This chapter won't solve all of your issues with time management. But we will try to outline a few strategies to help manage the demands on your time more effectively and efficiently. With good time management, you'll be able to accomplish more of those things that are important to you. One of the very first keys to good time management is understanding what is important.

If everything is important, then nothing is important

Since we all have limited time, good time managers know which of their responsibilities they need to spend their time on. A good time manager, first and foremost, understands what his or her priorities are. As a graduate researcher, you need to understand what tasks will help you in achieving your goal of a successful thesis and research.

Stephen Covey, in his classic text *Seven Habits of Highly Effective People* (Covey, 2004), introduces the time management matrix, which is a simple way of categorizing activities according to importance and urgency. To use the matrix, you need to identify all of the tasks and responsibilities related to your research, studies, personal life, and work life and determine if they are "important" or "not important" and "urgent" or "not urgent."

The difference between urgency and importance has to do with time (when does the task need to be done?) and goal achievement (does the task help me realize success?). Urgent tasks are those things that need to be done right away, while important tasks contribute to accomplishing your goals, whether research related, work related, or in your personal life. An example of a time management matrix for a graduate student is shown in Figure 8.1.

Creating the time management matrix and identifying tasks that reside in each of the quadrants is just the beginning in good time management. The idea behind the matrix is to reduce the size of Quadrants I, III, and IV, so that your matrix looks like the one shown in Figure 8.2, pushing you to spend most of your time on the important tasks and not on the urgent tasks. The tasks that reside in this quadrant are the things that will help you to achieve success as a graduate student and in other areas of your life.

Keep your focus on Quadrant II: Important, but not urgent tasks

A surprising thing happens when you organize your time around important, but not urgent tasks: your life becomes more fulfilling and less stressful. As a good time manager, you will find that you have more time to work on the things that you find fulfilling because you've identified what is important to you and are spending your time working toward those goals. This is not to say that ignoring the other quadrants is effective time

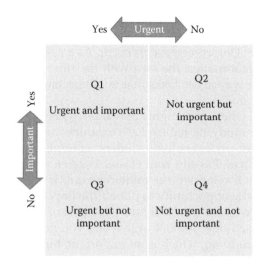

Figure 8.1 Time management matrix. (From Covey, S.R., *The 7 Habits of Highly Effective People: Restoring the Character Ethic*, Free Press, New York, 2004.)

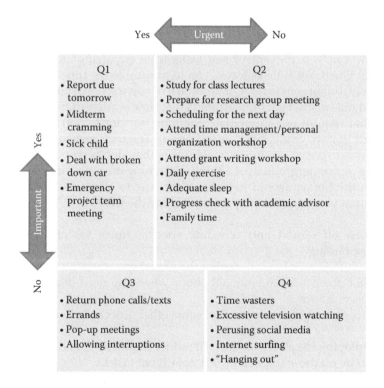

Figure 8.2 Example time management matrix for students.

management. Tasks will sometimes become urgent tasks (both important and not important), and you'll have to take care of them immediately. Surprises happen. Unforeseen events occur. As a responsible adult, you'll attend to the urgent matter; the idea with the time management matrix is that you have a way to see tasks that might become urgent and handle them before they do.

One example where you might apply this urgency mitigation strategy might be in your studying habits. Let's examine you as an undergraduate student. As an undergraduate student, cramming for exams might possibly be the norm. Passing your classes is given as an important goal. At the start of each semester, the midterm exam is several weeks away, and so you have the opportunity to put off the important study activities needed to prepare for the exam. But the early part of the semester is when these important activities are not urgent and you miss many opportunities to work on studying, which is an important, but not urgent task. As the exam looms closer, completing these study tasks becomes a priority not because their importance has increased (studying has always been important), but because they need to be accomplished before you sit down for the test (their urgency has changed). Now, days before the exam, you need to find time in your schedule to cram for the test which results in ignoring other important tasks such as exercise, sleep, and preparing for your other classes.

How can you prevent urgent tasks from occurring? Keeping your focus on Quadrant II tasks forces you to minimize the time spent on not important, but urgent tasks (Quadrant III, i.e., distractions) and not important and not urgent tasks (Quadrant IV, i.e., time wasters). Distractions are tasks that require our immediate attention but don't contribute toward the goals that we've identified as important. The trick to dealing with distractions is to create an environment where they don't creep in in the first place. In our undergraduate studying/cramming example, finding a quiet place in the library would have been one way to eliminate distractions. Tactics that you might employ might include the following:

- Turn off e-mail and schedule specific times for checking and responding
- Get off of social media during the workday
- Shut down/silence your cell phone when you need deep periods of concentration
- Close your office door or find some other quiet spot to work

Employing the above tactics will reduce distractions and allow you to concentrate on those important, but not urgent tasks.

Finally, time wasters are those tasks that occupy your time but do not contribute to your identified goals and there is no general reason for accomplishing them. You might enjoy these tasks, but they won't help you along the path to a successful thesis and eventual graduation. Just like the important, but not urgent tasks, you also need to identify time wasters. We provide a taxonomy of time wasters as a starting point for identifying your own set of time wasters to avoid and eliminate.

Cramming for exams is stressful (Quadrant I) and dealing with distractions (Quadrant III) and time wasters (Quadrant IV) might be fun, but spending your time on these things crowds the truly important tasks that reside in Quadrant II. With the right planning and personal organization, you'll find that you will have more time to work on the truly important things that will help you succeed in accomplishing your graduate education and research goals. But identifying tasks according to their urgency and importance is just the start. The following section outlines a simple strategy to keep your focus on the Quadrant II tasks.

A strategy for your time management

Separating your priorities from your time wasters and distractions is a good start. But time management is a control activity that must happen on a regular, day-to-day basis. Good project managers audit and review their budgets on a regular basis, and a good time management strategy is built around regular reviews. The following are some key steps to setting up a good strategy for managing your time.

Evaluate your current time management

You've done some of this already with the time management matrix. You've identified what is important to you and the activities that will help to achieve your goals. You also found tasks that you do every day that detract from achieving those goals. But taking a more detailed look at these tasks will tell you just how much time you spend in these areas and where you can make changes in your daily schedule to help with your time management.

For this, we recommend that you keep a simple activity log for a few days just to see how you are spending your time. The activity log is just as simple table with blocks for you to jot down how you spent a particular part of your day. In the log, you list the activity and the type of priority that activity holds. The log will help you evaluate the priority of the activity based on if it was urgent, important, or maybe even both (Figure 8.3).

Your review of the log should answer the following questions:

- What priority activities did I achieve? What priority activities did I not achieve?
- How much time did I spend on my priority activities? Is it enough?

Date:			
Time	Activity	Important	Urgent
7:00–8:00		☐	☐
8:00–9:00		☐	☐
9:00–10:00		☐	☐
10:00–11:00		☐	☐
11:00–Noon		☐	☐
Noon–1:00		☐	☐
1:00–2:00		☐	☐
2:00–3:00		☐	☐
3:00–4:00		☐	☐
4:00–5:00		☐	☐
5:00–6:00		☐	☐

Figure 8.3 Example daily calendar log.

- How much time did I spend on time wasters and distractions?
- Is there a particular time in the day that I am the most productive?
- Is there a particular time in the day that I am the least productive?
- How can I "chunk" my time according to high-priority tasks?
- Where can I fit in low-priority tasks?

This "postmortem" of a typical day is not unlike a project postmortem. In these reviews, your focus should be on what you did right with your time and what went wrong. You might find that you have good time management habits, but more likely you'll have areas that you can improve. But not to worry, your time management strategy is your own and is flexible to your personal and professional goals.

Develop a comprehensive calendar/personal organizational system

Time management is a very personal activity. The time management system that you employ to help with this should be as well. There are lots of great tools such as desktop-based apps, mobile apps, and even the traditional paper-based daily planner. Having a tool for your time management relieves you from the burden of having to remember all of your commitments. Without a system in place, you'll quickly forget tasks on your to-do list, stop thinking about long-range personal goals, and miss important commitments that you've made.

The system that you choose should be one that appeals to you because it will be a daily tool that you will use. Is your mobile device always at your side? If so, maybe a system based around a mobile app fits your style.

Are you always working on your computer? There are many well-integrated platforms, both free and commercially available. Or do you feel more confident with a paper-based system? Classic planners and time management kits are available everywhere. Making sure the time management system fits your personal style removes a barrier to using it and encourages daily planning of your time.

Make time to manage your time: The 30/10 rule

Speaking of daily time management planning, you need to make regular time to manage your time. The 30/10 rule is a good rule to follow for making time for time management. On a weekly basis, take 30 minutes and plan your week's activities. In this 30-minute weekly session, review your long-range goals and plan out when in the week to accomplish tasks which work toward these goals. On a daily basis, take 10 minutes every morning to review the week's plan and plan out the day ahead of you. These small and reoccurring blocks of planning time help in the regular maintenance of your time management system. The 30/10 rule is advantageous because it is quick and a regular activity that you do.

Identify and eliminate your major time wasters

We've given you a list of common time wasters and distractions in Table 8.1. But that table is just a starting point. From your time management evaluation, what are the time wasters and distractions that you are spending your time on? Can you possibly eliminate them altogether? Or at least reduce the amount of time that you spend on them? Consider the list that you come up with and ask if the tasks are really fulfilling and contribute toward your long-term goals and personal desires.

A simple test to see if you can eliminate these things on your list is to evaluate if the following sentence is true or false: "I really wish I would have done more of _____." If the sentence that you come up with is false, then the task is a solid time waster and a good candidate for eliminating from your time management plan.

Need to spend time at meetings?

Meetings are an important avenue for information flow for decision making. Whether the meeting is with your research advisor, class professor, or other classmates on a group project, effectively managing these meetings is an important skill for achieving your goals in a timely manner. Meeting attendees can often feel that meetings are avenues for wasting time and obstructing productivity, but this is because the meetings they attend are poorly organized, improperly managed, called at the wrong time, or just unnecessary.

Table 8.1 Taxonomy of time wasters

Self-imposed time wasters	Environmentally imposed time wasters
1. Over-socializing	1. Delays by others
2. Preoccupation	2. Unproductive environment
3. Attempting to handle too much	3. Unwelcome interruptions
4. Pursuit of perfection	4. Waiting for others' decisions
5. Inability to say no	5. Equipment failure
6. Lack of priorities	6. Others' mistakes
7. Procrastination	7. Lack of policies and procedures
8. Unrealistic time estimates	8. Lack of authority
9. Indecision	9. Lack of feedback
10. Self-indulgent distractions	10. Lack of communication
11. Lack of planning	11. Lack of cooperation
12. Too many mistakes/hasty work requiring repeats	12. Poor coordination

Consider the amount of waste that might occur by attending a poorly managed meeting. Every hour spent at that meeting is an hour wasted times the number of people in your group. From everyone's perspective, whether advisor, professor, or classmates, that hour could have been spent on some other productive task such as completing another assignment, reading, lab work, or working on one's research thesis. To prevent wasting time, we suggest the following guidelines for more effective meetings:

- *Do your premeeting homework.* Actions such as developing an agenda, sharing the agenda, listing desired outcomes, determining how to verify each outcome, determining who needs to be at the meeting, and finding a suitable place and time should all be performed prior to hosting the actual meeting.
- *Start the meeting on time.* All attendees should also arrive early. Nothing is more of a time waster than having to review progress and meeting notes to someone who arrives late.
- *Review the agenda at the beginning.* This will give all attendees a chance to have prepared responses and inputs to specific topics.
- *Get everyone involved.* Everyone present should have a reason for attending. If necessary, employ direct questions and eye contact.
- *Stick to the agenda.* Do not add new items unless absolutely essential. New topics will surely derail a well-planned meeting by taking precious time in unstructured discussion.
- *Be a facilitator for meeting discussions.* You will retain leadership and control of the meeting in this role. Be sure to quickly terminate

conflicts that develop from routine discussions and redirect any irrelevant discussions back to the topic of the meeting.
- *Recap the accomplishments of each topic before moving on.* This is an opportunity to remind those who have made commitments and articulate what is expected of them.
- *End the meeting on time.*
- *Prepare and distribute minutes.* Emphasize the outcome and success of the meeting.

We also recommend scheduling your meeting times strategically. When scheduling your week, block off times for several meetings back-to-back, or schedule meetings before your class times. This will not only force you to keep the meetings focused and end them on time, but will also allow you to have other large chunks of time during other days for activities that need to be accomplished in larger stretches, such as reading and writing.

First things first

Schedule the important things first. Whether in your weekly planning or daily planning sessions, making time for the high-priority tasks and scheduling low-priority tasks around them ensures that you have built space to accomplish those activities that are truly important. Studying, going to class, and researching/writing your thesis are examples of those activities that should be scheduled first. These are the important tasks that require large blocks of time in order to really accomplish the work associated with them. After blocking out this time, scheduling lower-priority, smaller tasks can be fit around these important things. Tasks such as errands and phone calls are examples of lower-priority things. Things that will take "just a minute" are examples of the low priorities that can be moved around easily in your schedule.

Also, don't forget about more basic needs such as sleep, eating, exercise, and spending time with your friends and family. These other high-priority tasks lay the foundation for you to be able to accomplish your work-related goals and tasks. Without making time for these "refueling" activities is a like driving your car with a near-empty tank: You might be able to go for a while, but eventually you'll run out of gas. Although very basic, they also deserve space in your time management system.

Odds and ends to contribute to your time management strategy

Finally, here are some odds and ends for you to consider as you develop your own time management strategy. Everyone's time is personal to them, and so your strategy for managing this time should be personal as well. Take what you want from the list below and come up with your own tips in personal time management. There's no need to try to implement everything; everyone will find their own usefulness in each of the tips below. Following simple and easy-to-remember rules will increase the likelihood that you'll implement a few activities for good time management.

- *30/10 Rule*: Good time management planning can take as little as 30 minutes a week and 10 minutes a day. Take this time to clarify your weekly/daily goals and objectives.
- *80/20 Rule*: Also known as the Pareto principle. 80% of your positive results will come from the vital 20% of your activities; 20% of total value is associated with 80% trivial pursuit; 80% of your time is consumed by 20% of value-adding activities. Use this rule to prioritize your activities.
- *Activity trap*: Avoid focusing on the activity or task itself and, rather, focus on the expected result from the activity.
- *Batching*: Batch small activities together (like e-mail). These batches can then be scheduled around important and major projects (see schedule block).
- *Cleanup and organize*: Always cleanup and organize before starting a major new task.
- *Clutter avoidance*: Avoid clutter so that you can clearly see what needs to be done and when.
- *Deadlines*: Set personal deadlines to inspire action and avoid procrastination. But never live and die by the deadline. Allow time for contingencies; last-minute pressure may be challenging, but is also bankrupting.
- *Easy-does-it*: Ease into tough tasks by warming up with simpler activities. Or break tough tasks into bite-sized chunks.
- *Formatting*: Establish and follow a standard format or rhythm of work that works well for you.
- *Grouping*: Combine similar tasks to facilitate concurrent achievement of results. A frequent mistake (and time waster) is to perform tasks serially.
- *Payoff hierarchy*: Start with the most profitable (highest payoff) or important parts of a task.
- *Prioritizing*: Classify your activities based on importance as opposed to urgency. Important things should take precedence over urgent things.
- *Planning*: Consistency in personal planning produces better results than no planning... and always in less time.
- *Productive mornings*: You are more productive in the morning than you are in the afternoon. By frontloading your day with important

tasks, you are taking advantage of higher energy levels and dedicating them to high-priority activities.

- *Schedule block*: Schedule an entire, reoccurring block of time for one major project. Tune out all trivial activities during that block.
- *Scheduling*: Scheduled events are more likely to happen than unscheduled events.
- *Time budgeting*: Allocate your time as if it is the most expensive resource you have.
- *Tooling*: Have a fixed place for each important tool. Return each tool to its rightful place promptly after use. This avoids misplacing the tools, forgetting where they are, and wasting time searching for them when they are next needed.

Reference

Covey, S.R. 2004. *The 7 Habits of Highly Effective People: Restoring the Character Ethic.* New York: Free Press.

chapter nine

On the personal aspects of research

In Chapter 2, we spoke a little about some personal aspects of graduate education such as evaluating your goals and desires and finding the right "fitting" program. This chapter returns to the topic of personal aspects, but from the viewpoint and perspective of a current student researcher working toward graduation. As faculty, we see students of all types with varying levels of success. This book attempts to provide a roadmap for success in graduate school, but just as each individual is unique, so is the graduate school and research experience. Therefore, there are different challenges that each individual student will face.

Student typology

In general, we can categorize students typically on a spectrum between class-oriented students and research-oriented students. Neither type is a good or bad student; this section merely points out and highlights the potential blind spots and strengths of each type. Since this is a continuum, you may fall somewhere in the middle—with attributes of both of these types. In that case, it is likely relevant to heed the tips and warnings for both of these types.

An orientation toward classwork: The classwork types

Many students find that achieving the near-term goals of course work to be rewarding and fulfilling experiences. After all, getting an "A" on the latest paper or exam and having accomplished another quarter of classes are tangible accomplishments that you can set your sights to or look back on. For students that are the "class-types," having these short-term, achievable goals are a concrete way to tell them that progress is being made (or not made) toward graduation. For the "class-types," their focus is on completing the tasks involved in classwork that are immediately achievable.

Undoubtedly, time spent working on classes is time spent away from research and thesis (i.e., unless the student is savvy and can combine efforts of classwork and research, a topic for later in this section). And so, the class-types may be less productive in his or her research area but only because the short-term motivators are not in place to get him or her working on the thesis. The seemingly open-endedness of the research project with its ill-defined goals may have created a mental barrier for the student to work on the research project.

The strengths of these types lie in their orientation toward the concrete and short-term. In particular, goal setting and detailed scheduling are skills that will help these students succeed in the classroom and can be transformed to help them in research. For them, classes are convenient "blocks" to plan around. Each class has its own set of objectives, milestones along the way, and even a predefined schedule, all of which are outlined in a course syllabus. These things are translated into the class-type's personal schedule, a notebook is easily created which is neatly segmented according to some organizational scheme, and, combined with other courses, the entire term is neatly laid out for the student. The goals for the quarter, as well as when they should be achieved, are known.

An orientation toward research: The research-types

On the other end of the spectrum are those students who are oriented toward their research. To these students, classes are understood to be an integral part of the graduate program, but they are different from the class-types in that they don't hold a singular focus toward classes and course work. Instead, their curiosity in a research idea drives them to be quite self-sufficient and self-directed in their research.

Like the class-types, research-type students have several blind spots to be aware of. Opposite from the class-types, they may well be disorganized in their planning of course work. This disorganization may not bother them because, after all, everything to do with classes is short-term and might be easily recoverable. Also, they may not see the point of a particular class or classes and so their performance in these "pointless" courses may be lower owing to a lack of motivation. Finally, research-types may be the kind of student that takes self-sufficiency and self-directedness to an extreme and not meet with their advisor for long stretches of time. This can be quite detrimental to that student's research if he or she is headed in some wayward direction.

The strengths of these students come from their innate curiosity which serves as their motivator for conducting research. Specific and concrete goals aren't needed for them. Instead, answering or solving some element related to their research or learning some new and interesting skill serves as the reward and fulfilling experience. Because accomplishing small elements of the thesis serve as individual rewards, these types might progress through the research process with less input and direction as compared to the class-types. For the research-types, time spent upfront in shaping the research question and methodology is time well spent.

Class-type versus research-types: Leveraging strengths and avoiding pitfalls

So how are your strengths as a class-type student or research-type student to be leveraged toward accomplishing your research? There are a few strategies we would recommend.

Leveraging strengths

For the class-types, you are very good at creating detailed goals and schedules on an academic term basis. Breaking your research down according to quarter (or semester), and then again, according to week will help in creating small blocks of time in which to set goals and deadlines for your research. A meeting at the beginning of each term for goal setting with your advisor will set up those tangible and concrete accomplishments that you can "check-off" as the quarter progresses. Should you begin to fall behind in your research, you can look back to this document and see what aspects you should put more attention toward.

One technique that will pay dividends for the class-type is to find a way to relate your class assignments and projects to your overall thesis effort. Think about ways that a particular deliverable in your class might contribute to your research question. How can the paper just assigned be used to address some aspect of your research question? Is there a literature search on a class topic that is related to your thesis that you can perform? Were there any analytical concepts and theories from class that you might further explore related to your research? If the relationship is tangential, look to see how you can adapt the assignment/paper/project to meet your research needs. Relating your class assignments (especially those that allow you to pick the topic) to your thesis effort helps you in spending time thinking about your research problem, forces you to do the hard work of writing down your ideas, and you get in-class credit for this research work!

Research-types will benefit from combining their classwork to their research, as well. For the research-types, however, their main strength is different from the class-types: they are able to conduct research in a self-sufficient, self-directed way. Be warned … only venture out on your own after you have a better footing in your chosen area. In order to get this footing, a few low-stakes activities might assist. First, a very early literature review is a low-risk, high-reward activity for you. This early review will help you get a sense of what the academic field says about your topic and without the pressure of a looming deadline or the uncertainty of the final research question. Another activity would be to discuss your research ideas early and often with faculty in your department. Finding the right advisor is an important step in the research process and discussing your

ideas with faculty will help generate more ideas, refine your current proposal, and help you screen potential advisors. For research-types, you
are motivated by an innate curiosity and interest in a particular topic.
Leveraging this motivation will provide you the energy needed for your
research. You'll find that you are genuinely interested in your thesis topic
and by keeping your research interesting, you'll find that doing the work
required for research will be less like work and more like play.

The pitfalls and how to avoid them

With strategies for leveraging strengths, we also suggest some strategies to help you spot potential pitfalls and blind spots as you conduct
your work. One pitfall for the class-types is the tendency to delay their
research work in favor of course work. One reason for this could be that
the class-types might find the entire process of research bewildering
and overwhelming because of the open-endedness of the entire process.
Unlike classwork, there is no "correct" solution in research. Without a
clear set of instructions or a defined end-state, you are expected to accomplish a bewilderingly massive project. And in the end, you will be faced
with convincing a research committee that your problem, method, and
analysis are sound and that the results and conclusions are valid. Only
then, after meeting seemingly arbitrary standards, will your work pass
as thesis research. With a lack of a concrete set of goals, the class-types
might have a tendency to put off the research work, choosing instead to
focus on their classwork. The immediacy of assignment feedback and
the end of the term signifying the closure of a stage in their academics,
classwork is more tangible and gratifying in the short term. Classwork is
an important part of research because it lays the foundation necessary for
the exploration, but too much time and attention in this area only delays
the hard work needed to solve the selected research problem. Leveraging
the strength outlined in the previous section such as scheduling research
in small blocks of time and combining classwork with research work will
help mitigate some of the procrastination that results from trading off
research time for class time.

Another pitfall that the class-type might have is the need to turn in
a completed product for their advisor to review. Completeness in your
writing does matter; but it matters in different degrees at different stages
of the research. At the very beginning, initial thoughts, notes, and very
rough drafts are things that are completely acceptable. Only the most
unreasonable of advisors would expect perfectly formatted work from his
student. Early on in your research, getting your ideas onto paper prevails
over the need to have perfectly structured and formatted work. Substance
matters, but you need to be able to share that substance with your advisor. As you progress in your program, and surely toward defense, the

substance of your work will become well defined, and structure and formatting will take priority as you prepare the final document and deliverables. For early deliverables, begin by using outlines and then progressing toward story-boarded slideshow presentations. This will allow you and your advisor to focus on content, without the distraction of not having a polished final manuscript.

With research-types, the pitfalls you need to watch out for are very different. You might revel in the unstructured, open-endedness of research, but because of this, you might run into problems of scope and schedule. Changes in the scope of your problem are intimately related to changes in the accompanying schedule. Scope creep is a project management term for when tasks or outcomes of a project are added, either by members of a project team or by the project's stakeholders. In general, things are added to the scope in order to increase functionality, features, or capacity for the final product of a project. Similarly, in research, scope creep adds features to the research project in the form of different or new research questions, additional data collection, and additional analysis. Although interesting and relevant, the danger for a research-type is that the project becomes a never-ending story and a final completion date never materializes.

To mitigate this problem, having a well-defined research question limits the area of research and helps to prevent this scope creep. Consider a few of these methods below.

- *Answer the question: "What is your research question and why is it important?"* Try to do so in a succinct two to four sentences. With a succinct response, chances are that your question is in fact well defined. If you find yourself rambling, your scope might be too broad.
- *Create a 30-second elevator speech.* By being able to distill your research in such a way to be complete and understandable to a stranger, you pass a different test of succinctness.
- *Test your topic.* With the resources available to you in the library, see if you can find any information on the topic and the question. What you find (too much information, too little, someone has already answered the question) will tell you if you need to rescope your research.
- *Talk with your advisor.* But first, write down what you think your research question(s) should be. Your advisor will know if your problem is properly scoped given the time frame for your program.

Very intimately tied to scope creep is the problem of staying on schedule. Additional tasks added to the project, without consideration for adding additional resources, will only add more time for project completion. Developing a schedule, constantly reviewing the schedule, and making needed adjustments to the schedule will help in meeting the overall goal on on-time graduation. Understanding how changes in your research question fit into the existing tasks will help you in understanding the impact to milestone dates in your research and the overall impact to the final thesis defense and graduation date.

We have two useful tactics to employ regarding your research schedule. First, you should schedule regular reviews of the schedule with your advisor. The regular reviews ought to be conducted once a term. Consider combining these reviews with a list of goals accomplished from the previous quarter and goals planned for the next quarter. This goals and schedule review will serve to be a productive meeting as you demonstrate the accomplishments made so far and discuss near-term planning. Second, consider placing your schedule somewhere visible as motivation for the work ahead of you. By doing so, you not only see what is ahead of you, but you're also reminded of the progress you've made thus far. As you progress in your program, you'll be able to quickly see if the posted schedule is a reflection of reality; if it is not, that's your cue that the schedule needs to be reworked and rediscussed with your advisor.

Knowing which type of student you are will help in identifying your strengths and weaknesses in different areas of your graduate education and research. Knowing your own personal attributes identifies your predetermined preferences for the type of work ahead of you. If you know that you do not like to do detail-oriented tasks such as planning and scheduling, chances are that you are not very good at them. Knowing what strengths and weaknesses you have, and understanding that these come from a predetermined preference of the type of student that you are, will help you see what areas you might want to spend time developing these skills.

Personal skills and qualities

So we've spent some time putting you into a box: maybe you're a class-type student or maybe you're a research-type student. Regardless of where you fall, understand that none of the skills are inherent traits that can't be further developed. Rather, you can spend time taking (another) class or reading a personal organization book (like this one) to help you develop the necessary skills for successful education and research. There are other skills that we'll call "soft skills" that are just as important as skills you learn in a classroom or other formal training. These soft skills are qualities that you can pick up through self-reflection and self-improvement.

Why are soft skills important? Soft skills are traits and skills that define how you deal and interact with the people around you. Without these skills, you simply won't achieve the same levels of success in graduate school (and in future jobs) as compared to those with these skills. What are these soft skills? We have listed some below and describe how they relate to your research and graduate education.

- *Have a positive attitude.* Research will get tough. You will experience a lot of uncertainty and maybe doubt your ability to finish, but a positive outlook will help you through the tough parts of your education. A positive outlook will translate to energy and enthusiasm toward your work.
- *What's your motivation?* In order to achieve this right attitude, try to understand why you are in graduate school. What are your goals and desires in the short term (i.e., this semester)? In one to two years (i.e., before you graduate)? And in the long term (i.e., 4 to 5 years)? After figuring out your goals, assess if the work that you are doing today relates to your goals and desires. Making this connection is important because you'll be able to see how today's work contributes to your goals and desires. And if you can see the connection, you're more likely to develop the discipline to make it through your program. The best students are driven: They know what direction they are headed and are working every day to get there.
- *Collaborate with your classmates.* Learning to work with others is an essential skill you'll need in the workforce after graduating. Thankfully, graduate school has many opportunities for collaboration whether in your classes (homework, projects, study sessions) or in extracurricular activities on campus. Collaborating helps in building relationships with your peers that you can turn to for different areas of help. A strong network of peers will be the group of people that you can turn to for sharing ideas, seeking feedback on these ideas, asking for help in areas where you lack skills, or simply having a group of people that you are accountable to for daily, weekly, and monthly progress.
- *Maintain your personal relationships as well.* In your collaborative network, you may form some personal friendships and you'll also have your relationships from before graduate school. Friends and peers that you trust will be there to help you through the tough times in your graduate education. You'll also find that whatever insecurities and anxiousness you might have about your progress isn't unique to you. Simply sharing these feelings will go a long way in boosting your personal morale. Or you might be the one offering advice to someone going through their own difficulties. Whether receiving or giving advice, you'll be surprised at how much influence the personal relationships you foster have on your success in graduate school.
- *Take the initiative in your own research work.* Having a strong network of good people around is essential for success, but being able to forge

your own path is just as critical. An ability to self-manage, act independently, and take the initiative goes a long way as many aspects of research are solitary endeavors. Writing your thesis, thinking about the research, and planning your next steps are just a few tasks that you might find you'll need to do on your own. You shouldn't wait for your advisor to assign you specific tasks to accomplish; in fact, he or she will appreciate you having taken early action with your research work. If you don't quite know what you are doing, that's okay, just get started because a good research advisor will be there to mentor and guide you through the process.

- *Have self-confidence because it will carry you a long way.* So you've thought about taking initiative, but you can't quite get started because you don't quite know what you are doing (see above). If you have ever thought that maybe you don't belong in graduate school because you don't know what you are doing in your research, don't worry … those thoughts are completely normal. Everyone had to start somewhere, including your research advisor and the straight-A student in your class. Dwelling on the feeling of not belonging will be counterproductive to your work and will only serve to damage your self-esteem further. Remember that school is a time for you to learn and that it's okay that you do not know as much as you feel you should. You have the opportunity to learn and work at new skills and in a new field. A high sense of self-confidence will allow you to try new things, cope with the stresses of your graduate research, relate to people better, and overcome many of the difficulties that you might come across. Self-confidence is a positive feedback loop: act with self-confidence, people will treat you as a confident person, and this positive feedback will build more self-confidence.

- *Grit: The stuff of resilient character.* Grit is the ability for you to dig down deep for the motivation to finish, carrying on with your work even in the tough times, and having the courage to stick with the program when things look bleak. In all of these soft skills that we've listed, the ability to persevere might be the most important skill to try to develop. The road to graduation will seem long and hard, especially in the middle years. In the middle years, the honeymoon period of a new environment, novel research, exciting new people, is gone; but you are nowhere close to the emotional highs of nearing graduation, nearing the finishing line. The middle years are where many graduate researchers' insecurities about their work, anxiety about finishing their research, and maybe even the loneliness of being a researcher will rear their ugly heads. Grit is the courage to show up to work every day and resolve to accomplish some task on your research schedule. A resilient character will help you in coping and managing your stress through these tough times.

Grit, tenacity, determination, resilience, or resolve—whatever you call it—buckling down and working hard will get you to success in your graduate program.

Dealing with the stress of research

Undoubtedly, things will go wrong (or at least feel like they are going wrong) in your program. Some of the classes may be more difficult than you were expecting, you found that your research idea is not as novel or as easily researched as you thought, you do not have as much time at home to devote to study and work, or maybe you've realized you are not going to hit this semester's goals which will significantly delay your graduation. All of these things going wrong will cause you stress and you must identify and deal with stress head on, so that you can ensure success in your graduate program.

Graduate research can be a stressful time for any number of reasons. If you are not working, you might not be bringing in much of a salary and are facing some financial difficulties. Or you might be working and find that, because your time is occupied with school, you are neglecting important elements of your life—such as friends and family. Some of the stress might be induced from your research work: maybe you are falling behind schedule, maybe your advisor is not giving you as much feedback as you hoped, maybe you feel that your advisor doesn't care about your progress, or maybe you are simply unsure of the direction of your work.

Any of these problems can lead to what we'll call the "research blues." Research blues is simply the lack of motivation to continue working on your project. For PhDs, this might occur during the middle years of your program where you no longer have classes with their short-term, motivating goals. For master's students, this might fall somewhere between the problem definition stage and the data collection stage of your research. In either case, you are somewhere in the middle stages of your research when it seems as if you have nothing to show for your work. Falling into the research blues is very easy. The days become a grind that is neither enjoyable nor productive. The lack of motivation leads to a lack of progress, which in time will lead to stress as you try to figure out if you need to leave the program or you fear that failure in the research is inevitable. These feelings are all very common for students and understanding what you are feeling is the first step in coping.

What are the classical symptoms of stress?

- Tension headache
- Muscle tension
- Anxiety
- Self-doubt

- Constant fatigue
- Nervousness
- Paranoia
- Increased heart rate
- Increased blood pressure

These classical symptoms are the reactions that your body exhibits when encountering a stressor, whether from graduate school or from your personal and work life. Although these are natural reactions, experiencing these stressors shouldn't occur for prolonged periods of time. As a student, you might be tempted to push yourself to the limit in order to squeeze a few more hours of work out of your mind and body. Case in point: all-nighters. During your undergraduate years, pulling all-night study sessions to cram may have gotten you through your classes; but graduate school is a different time. You most likely have responsibilities other than yourself and you can't afford the time to "crash" afterwards.

Extended periods of stress will likely cause some sort of physical or emotional problem in that person. A person that allows stress to be present for extended periods will experience burnout, depression, or illness unless he or she takes some action to manage and cope with the stressors. Completely eliminating stress is unlikely; in fact, some levels of stress are necessary to achieve high performance. So the only way to manage stress is to delicately perform a work–life balancing act. Here are some tips to perform this balance:

- *Create and maintain a support structure.* Surround yourself with good people to talk through problems and reach out for advice. You'll be surprised at how much relief you'll feel just from talking out your problems and knowing you have people that will help you.
- *Relax.* Create opportunities when you can relax your mind and body. Taking short breaks during the day will help clear your mind; taking time during the weekend or after school will help you be fresh for the next day's challenges.
- *Set small manageable goals.* Achieving a predefined goal provides you with a sense of accomplishment and with this accomplishment, some stress relief. Try not to overextend yourself. If you have a large task in front of you, 30 minutes of planning time to strategize and create small goals will go a long way in reducing stress.
- *Get sleep.* Sleep is a basic need for everyone. Inadequate sleep diminishes your cognitive performance and inadequate sleep for extended periods is detrimental to your overall health.
- *Take time to eat.* Take time to eat breakfast every day and proper meals for lunch and dinner. Just like sleep, this is a basic need which gives you the fuel to perform. Don't settle for snacks that do little in

providing you nutrition. Find the right foods to eat and you'll find that this makes a positive impact on your performance at school.

- *Exercise.* Just like finding time to relax your mind and body, exercise is a good way to clear your mind and refresh you for the tasks ahead. This diversion will give you a renewed sense of vigor when you return to the research tasks in front of you.
- *Humor me (yourself).* Use humor to cope with the demands of research. Laughter is a great form of stress relief.

Dealing with failure

We hate to end the chapter on a negative note, but failure in your graduate research program will be inevitable. Failure in research lurks at every turn: your conference paper was rejected, you didn't pass that all-important mid-term or your qualifying exam, you were asked to repeat your prospectus defense, or the data from your experiment were not close to what you were expecting. Regardless of the cause, you will experience failure multiple times in your research program. Scientific knowledge is built upon a series of failures. It is how the scientific method works, and it will happen when you conduct scientific research too.

When failure happens, you simply need to move past it. You should pick up the pieces of your problem and look to see where you can go from there. What can you learn from this experience? How can you improve your experiment, presentation, or conference paper as a result of the letdown? Why did it happen?

You'll need to set aside some time to separate yourself from the failure to be able to think about what happened with a clear head. Sleeping on it and waiting for the light of the next day might be a good start, but maybe you might need more time. Whatever the length of time, be sure to address the problems and surely do not ignore or let the underlying causes fester.

Also lean on your support network to help you get past your difficult experience. Everyone, and we mean everyone, has failed at some point in their lives. Talking with others will make you feel better about the situation, help you with reflecting on the experience, and give you a different perspective on the issue.

If everyone has failed, then how is it that everyone is not a failure? The difference is in how each individual handles their own failings. The successes have learned from their mistakes, moved on, and are better persons for having the experience. They didn't let the failure define who they are; instead, they allowed the failure to shape and transform them into a better individual.

Remember, everyone will have to deal with failure; the successful ones learn from their failures and are better for it.

chapter ten

Managing your research advisor

Dealing with or interacting with an advisor is very much like any other human-to-human interpersonal relationship. What you see in normal human relations is what you will see in your relationship with your advisor. For this reason, this chapter uses the theme of "managing your research advisor." Thus, we place you in the driver's seat of ensuring a successful relationship with your research advisor.

Effective student–advisor consultation

Soon after enrollment in a graduate program, you should confer with the director of graduate programs in your department concerning the plan of study. The director will assist you in preparing an initial advisory conference or will direct you to a faculty member for assistance. If a research advisor has already been selected or appointed for you by the department, then you should initiate all your consultations with that advisor.

You must recognize that graduate faculty members have diversified responsibilities for teaching, conducting research, writing journal papers, advising students, mentoring graduate students, and providing professional service. Understanding these varied responsibilities will help you develop a rapport with the faculty. Do not occupy the advisor's time needlessly. In your interactions with your advisor (and other colleagues), respect yourself so that others may respect you in return. Because of the varied nature of their scholarly work, graduate faculty may, at first, appear withdrawn, nonchalant, or uninterested. But upon closer examination, you will find that all they do are, indeed, designed to support your graduate study objectives. You should help them as much as they are expected to help you. To develop a productive working relationship with the graduate faculty, particularly, your research advisor, you should practice the following:

- Keep in touch with the faculty, but don't pester them.
- Respect the busy schedule of the faculty.
- Be prepared for initial self-help before consulting faculty.
- Most minor questions can be answered by available published guidelines.
- Don't drop in indiscriminately.
- Take the initiative for your graduate study and research.

Resolving conflicts

Conflicts develop in any human relationship whether personal, formal, or official. How you deal with the conflict will determine the outcome. Communication is often at the root of conflicts in any personal or professional engagement. This is particularly true in any research environment. Figure 10.1 illustrates the importance of communication with your advisor. The solid-filled circle representing you is used to indicate your own resolve and definite pursuits of your research goals. The grayed filled circle for the advisor represents the potentially vague or less committed enthusiasm of the research advisor. The patterned filled circle, for the research process, represents the volatility of research processes, requirements, and institutional guidance. This scenario puts the onus on you to ensure that proper communication and coordination happen throughout your research experience. With proper communication comes more reliable cooperation from the research advisor. The next sections present some guidance on how to handle specific conflict situations.

My advisor's instructions are too vague. How do I deal with this?

You should realize that not all conflicts are explicit or obvious at the outset. There are subtle misunderstandings that can manifest themselves as conflicts later on. If your advisor's instructions are too vague, you can remedy or mediate the situation with more proactive questions to the research advisor for the purpose of eliciting more structured instructions from him or her. To facilitate your effective questions, you should take extra steps to be more familiar with institutional requirements and guidelines. Most advisors are not conversant with the changing landscape of institutional requirements. After all, they are not the ones pursuing a graduate

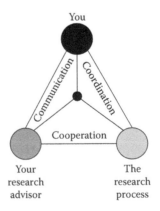

Figure 10.1 Communication loop for research management.

degree. Under the pressure of turning out more scholarship outputs (e.g., journal publications), the task of reading institutional requirements will take on a lower priority. So, research advisors will hardly take any extra time to learn and keep up with institutional requirements. This is particularly more common with long-term and experienced advisors, who might have been familiar with the distant-past requirements and still believe that the old and outdated requirements are still valid. With you learning more on your own, you can guide and lead your research advisor to the less ambiguous instructions that you need to complete your research.

How do I manage my advisor's expectations of me?

An advisor's expectations can, indeed, go awry and become unaligned with your own research goals and career objectives. To manage your advisor's expectation of you, you must first ensure that the advisor's expectations are valid and progressive for your research advancement. Again, we suggest you to implement the SMART (specific, measurable, aligned, realistic, and timed) assessment approach. In communicating with your advisor, help him or her to be specific with his or her expectations. This can be done with pointed, but courteous, questions to the advisor. Then, discuss with the advisor on the need to make the expectations measurable. Convince the advisor that a measurable expectation actually makes the advisor's job easier later on in terms of assessing (i.e., grading) the expectation. Note that a specific and measurable expectation still has the potential of being unaligned. So, you need to tactfully negotiate with the advisor to ensure that a specific expectation is properly aligned with what you and the advisor expect to see as the end result. To do this, remind the advisor of the previously approved research topic and the expected contributions. If you phrase your communication in terms of the advisor's own research interests and goals, you will have a higher chance of eliciting aligned expectations from your advisor. Following this, you and the advisor need to communicate on the realistic aspect of the expectation. If an expectation is specific, measurable, and aligned, it may still have the potential to be too grandiose for actual accomplishment. So, ensuring that an expectation is realistic at the beginning will save you some agony later on. Finally, be sure that you get the advisor to time the expectations. An untimed expectation can become subject to scope creep, which can make you not graduate on time. Once all the SMART elements of the advisor's expectations are assured, your next task is to use the principles of project management, as provided earlier, to manage, track, and control the expectations. In this regard, consider yourself as the single resource available to allocate to the several tasks associated with meeting the expectations. You must plan, organize, schedule, and control yourself and your project using disciplined project management techniques.

What should I expect from my advisor?
How do I communicate my expectations?

What you should expect from your advisor is a general guidance on what to do. You will have to fill the void in whatever the advisor does not provide. One strict expectation for graduate studies is the ability to do independent research, think critically, and communicate effectively. The advisor cannot do these for you. So, don't expect him or her to deliver the "holy grail" for you. Using the SMART approach presented earlier, you can communicate your own expectations to your advisor with the goal of having a melding of minds. If you communicate your own expectations clearly, you are more likely to get SMART expectations from your advisor in return. Remember that graduate research is a partnership between you and the advisor. Although the advisor has more research experience, you bring something equally important to the team. That is your exuberance and inquisitiveness about the research and the eventual outcome. Be sure to communicate your readiness to learn and work hard under the guidance of the advisor. Some of the suggested types of communication that you can use with your advisor are listed below:

- Formal written
  - Plans, details of complex problems, requirements, charters, high-level meeting notes, reports
- Informal written
  - Notes, e-mails, simple meeting notes, memos
- Formal verbal
  - Presentations to groups, speeches, presentations to research advisors and collaborators
- Informal verbal
  - General meetings such as periodic status updates, general conversations

We recommend you to take advantage of each communication type depending on the prevailing scenario. Sometimes, you can mix and match communication types, particularly if one is used as a follow-up to another.

My advisor is really busy. What do I do if my
advisor doesn't have enough time for me?

All advisors are busy, but that should not be an excuse. An advisor enters the "contract" of guiding your research with the understanding that adequate time will be committed to the research partnership. With the project management skills you learn in this book, you can actually help your advisor to be more effective and efficient with his or her own time-based activities. If you don't needlessly occupy the advisor's time, then you are helping to free up time for more productive engagement with your research. You should let the advisor know that you are sensitive to his or her time demands and you want to be less demanding of his or her time. If, in your own mind, the advisor does not have enough time for you, you should strive to get the most out of whatever little time you get to spend with the advisor. Note that what you term as "not enough time" may be viewed as perfectly sufficient time by the advisor, under the principle of you doing independent research.

What do I do if I am not getting timely feedbacks?

Not all advisors are good project managers. Timely actions are the hallmarks of project management. Unfortunately, those not practicing project management techniques often falter in timely actions. If you are not getting timely feedback from your advisor, you should use your own project management skills to elicit timely responses from him or her. If you do your planning well, you can stimulate prompt actions from your advisor by apprising him or her of the timeline and milestones in your research plan. If you put the milestones within the context of the advisor's own timeline for research productivity, you will likely see an upswing in the timing of the advisor's feedback. You can also influence the advisor's responsiveness by arranging regularly scheduled research advising sessions. Again, if you practice the principle of SMART, as presented earlier, the timing of feedback can be improved because the advisor will have a better grasp of when and why prompt feedback are essential.

What do I do if the feedback I am getting is not sufficient?

If your advisor's feedback is not sufficient for you to make satisfactory progress with your research, our advice to you is to take control of the situation yourself. Do not fall prey to resignation or indignation. Rather, take control by requesting specific response to specific questions that you pose to your advisor. Do not ask generic questions that could lead to generalized feedback. In this regard, if you conduct your research in decomposed phases, you can guide your advisor to address specific components of the research.

Is my research topic growing? How do I fend off new requirements?

The gradual growth of a project is called as scope creep, which is the addition of requirements and functionality to a project without addressing the effects on time, costs, and resources, or without prior agreement with the project executor. Scope creep is one of the dangers of improper project management. We encourage you to use the SMART approach, presented earlier, to stem scope creep. To avoid uncontrolled growth of your research, work with your advisor earlier on to set realistic boundaries for your research. If it appears that the boundary is being violated, consult with your advisor directly and rein things back into the approved research focus. Non-value-adding growth of a research topic can be stemmed by proper topic planning from the beginning. You and your advisor should prepare a clear statement of the topic at the beginning. This will ensure that your research topic meets the following requirements:

- Defined
- Verified
- Agreed upon
- Traceable and controllable
- Divided into discernible and trackable elements

What do I do if my advisor keeps changing my topic?

Erratic and incessant changing of topics is a bad sign of the advisor's own lack of research advising competence. Once again, the "specific" aspect of the SMART approach can ensure a commitment to the specific line of research that you and your advisor agreed upon. You can explain to your advisor that a research topic that appears not to be leading to the intended end result can still generate contributions in terms of the creation of new knowledge. A documented failure may make a contribution by veering others from following the same line of research. Having a documented statement of topic at the beginning can help you avoid your advisor's tendency to change your topic. Also, demonstrating a measurable progress with your topic is a good way to convince your advisor that you are on the right topic.

How do I tell my advisor that I can't finish my work on the current schedule?

If the schedule is realistic and achievable in the first place, temporary delays should not present an insurmountable obstacle. Frank openness and full disclosure are essential for achieving credibility with your advisor. Once you have credibility, it would be easier to present the actual status of your schedule to your advisor without appearing to be presenting a frivolous excuse. Good research topics will, undoubtedly, present challenges. If research was easy, everyone would be doing it. Your experienced research advisor would be able to recognize that challenges and unexpected delays are common elements of a good research topic. The key is for you to demonstrate that you are dedicated to overcoming the challenges of the research. If you keep your focus on the end result of your research, you would be able to leverage incremental achievements to reach the end result.

What do I do if my advisor is causing delays in my schedule?

Before you pick an advisor, you should be aware of the prospective advisors' reputation for responsiveness. If a delay does not represent a repeated pattern by the advisor, everything should work out fine in the end. But if the advisor is repeatedly delay-prone, you will need to take a more proactive stance in driving the research train. You should consult with the advisor on a more frequent basis as a way to get him or her to be more sensitive to your research schedule. A long lead time between student–advisor consultation means that each document for the advisor to review would be more voluminous, which would require more review time on the part of the advisor. Providing smaller chunks of review materials is a good way for you to ensure a faster response time from your advisor. You never want to overwhelm your advisor with a large document to review and appraise on a short turnaround timetable.

How do I ask for resources?

Ask, you might receive. Never feel intimidated or bashful to ask for whatever additional resources you need to get your research done. If you have kept true to the spirit of your research, you should be conversant with what each element of research needs in terms of resources. Your job is to adequately communicate those needs either to your advisor or to the leadership of your academic department. Your primary job as a researcher is to be the most ardent proponent of your research. This requires you to communicate and sell your research to all constituents. In this regard, you should be bold enough not only to brag about your research accomplishments, but also to plead your case for more resources to further advance your accomplishments.

How do I manage conflicts between committee members and/or my advisor?

Students can easily get wedged in-between feuding committee members. If you find yourself in such a situation, never take sides, even with your advisor. The best you can do is to help each committee member focus on the research itself rather than the personalities involved. Committee disagreement and conflicts often have their roots in other previous discordant interactions, to which you may not be privy. One reason that conflicting members may end up on the same committee could be the fact that there are limited other options available for the research topic at hand. As the person who has the most to gain from committee harmony, you should take it upon yourself to serve as the melding point for the warring factions. If unruly discussions erupt during your committee's meeting in your presence, we advise you to respectfully refocus everyone's attention to your research topic. Make yourself the focus of attention by diverting each person's conversation back to you and your exciting research.

When do I need to find a new advisor?

When all else fails, be prepared to bail out. If it appears that a sour situation cannot be salvaged, it is better to end the advisor relationship earlier than later. It is your research, it is your career pathway, and it is your future that is at stake. So, don't worry about hurting the advisor's feeling. He or she is, probably, as glad to disengage as you are. Do not subject yourself to the internal politics of the department when searching for an alternate advisor. Once again, focus on your research and the best person to help you reach your end goal. How do you know when the time is right to change your advisor? You will feel it in your inner gut. It is time to change when you are convinced the advisor is no longer able to meet your research advancement goals. You have a limited time to accomplish your research goal. Each wasted week is a significant impediment to your research completion. If you have gone four solid weeks without a measurable research accomplishment and it is not due to your own doing, then it is the right time to make a change. Before you announce your intention to your advisor, be sure you have consulted with a potential more compatible faculty member. You should also consult with the director of your department's graduate programs to ensure you can comply with the requirements and the process for changing research advisors.

Conclusion

The onus is on you to manage your research. This requires managing not only your research tasks, but also managing those involved in the research, particularly your research advisor. You should be a stickler for personal responsibility. You must demonstrate a whole-person capability covering a variety of skills and personal attributes. Some of these include the following:

- Time consciousness
- Interpersonal etiquette
- Taking personal responsibility
- Effective communication
- Adequate preparation for class

- Good study habits
- Compliance with institutional requirements
- Conformance with social norms and standards
- Contingency planning with respect to "Murphy's law"
- Learning to allow sufficient lead time between consecutive activities
- Planning, scheduling, and controlling study-related activities
- Managing independent assignments
- Familiarization with university resources, such as libraries, labs, etc.
- Positive interactions with other students

Most graduate students already have a good level of academic preparation. Where they need help with is being organized and managing themselves well. The essential characteristics for success include the following:

- Intelligence
- Ingenuity
- Creativity
- Self-organization and management
    Organize, organize, organize, and organize again!
- Knowing where everything is kept to minimize non-value-adding futile search for items

Knowledge and smartness alone are not enough to succeed in a graduate program. You must have the skills necessary to apply the knowledge and the attitude required to execute your research successfully. If you can manage the interactions with your advisor, you can control the success of your research.

section five

Project phase-out: When is research complete?

chapter eleven

Communicating your work

> I have not failed. I've just found 10,000 ways that won't work.

Thomas Edison

Of what use is your work if no one knows about it or appreciates it. You must communicate your work in a way that not only gets the attention of your audience, but also imparts new knowledge to the audience. Your audience will appreciate the quality and value of your work based on how well you communicate the work. For oral presentation in person, your audience can range from one person (e.g., your research advisor), a few people (e.g., your graduate committee), or a classroom (e.g., your co-students) to a very large group (e.g., a technical conference group). For written presentation (i.e., publication), your audience can be equally varied and diverse. In each case, you must adapt your communication style, depth, detail, and length to the specific needs and requirements of the audience. A published communication can be particularly challenging because you don't have a face-to-face rapport with the audience. For this reason, your published work must be designed to "speak" for you. Be sure to follow and comply with all the guidelines provided for the particular type of communication that you are doing.

Whatever the nature of the outcome or result of your research, you must still communicate it as effectively as any other type of information dissemination. Even a less-than-stellar result or a failed experiment must still be communicated effectively. What does not work can be communicated in a way that diverts the efforts of others from the same failed path. The opening quote above suggests that there is success embedded in communicating failure. Thomas Edison believed there was success in proving the thousands of ideas that didn't work for him. There is a tendency on the part of graduate students to shy away from reporting what did not work. This book's lesson to you is to communicate everything that you have done. In a quote by the coauthor Badiru, "if a work has not been communicated, it has not been done." For others to know what you have contributed, you must communicate your work. If anything at all, your failure can make a contribution in the sense of veering others from following the same path of failure. The following addition quotes provide a motivation for reporting everything:

> Just because something doesn't do what you planned it to do doesn't mean it's useless.
>
> **Thomas Edison**

> There's a way to do it better—find it.
>
> **Thomas Edison**

In research and academia, the most sustainable avenue of communication is archival publications. Oral presentations at conferences and in-person briefings are essential only as long as they pave the way for a journal publication afterward. The size of the audience in presentation is limited, but the size of the audience for a published work is innumerable. You will not always have an opportunity to speak in person directly with those who should know about your work. Thus, published works expand your horizon of intellectual visibility.

Understand that a good idea poorly expressed will sound like a poor idea. So, you must dedicate adequate time and effort to expressing and communicating your work effectively.

The quality of the presentation of your research results, both in writing and orally, will determine how the audience perceives the work. This chapter presents guidelines for effective technical reports and presentations.

Ordinary report versus technical paper

There is a subtle difference between an ordinary report and a research paper. An ordinary report involves finding and recording facts already available somewhere. A research paper involves not only finding the facts, but also evaluating, interpreting, and adding to the facts. While a research paper may require you to have your own hypothesis about the topic, a report merely requires you to represent the facts. A report may be comprehensive, covering various related ideas and topics. A research paper, by contrast, requires a specific and sufficiently narrow topic. Be very judicious in selecting the title for your work. Yes, you can quickly judge a work by its title, just as you can, indeed, judge a book by its cover. The following titles illustrate the difference between a report and a research paper:

Report Title: "Applications of Learning Curves in Business."

Research Paper Title: "Extensions to the Log-Linear Learning Curve for Cost Control."

Report Title: "Project Management in Industry."

Research Paper Title: "Resource-Constrained Scheduling Heuristics in Multi-Product Assembly Lines."

A common mistake by students is to want to put everything imaginable into their reports. It turns out that many of the items contained in student reports are irrelevant "fillers." A part of the training of graduate students should involve the development of the maturity to be concise in technical reports. Putting too much in a write-up is often a sign of lack of confidence that an adequate job has been done. The elements for an effective research paper include choosing the right topic, developing a succinct, representative, and informative title, collecting and evaluating reference materials, organizing ideas, developing an outline in accordance with specifications, and writing, editing, refining, and proofreading the draft.

Guidelines for technical reports

In general, your report should cover the background, objective, methodology, results, and contribution. Improve both the aesthetics and contents of the report by following the guidelines below.

Suggested report format

1. You should have a cover page clearly showing the title, your name, date, and topic.
2. Include an abstract that describes the problem, methodology, results, and conclusion.
3. Follow any specific requirements for the report. For example, most reports require a double-spaced abstract and text, and single-spaced reference list with a space between entries.
4. Journal article references should contain the name of author(s), year of publication, article title (in quotes), journal name (in italics, boldface, or underlined), volume number, issue number, and inclusive pages. Use any of the standard reference formats available and be consistent with them. Whenever a required format is specified by a journal, be sure to use the prescribed format at the outset. This will save you time with reformatting later on.
5. For long reports, include a table of contents.
6. The first two sections typically should cover an introduction and a literature review. The introduction should describe the general background for the problem. A subsection of it may present an explicit problem statement. The literature review should discuss what has been done before (as documented in the literature) and how your work differs from or adds to the existing work.

7. Be sure to explicitly and frequently recount the contributions of your work.
8. For ease of reference, always paginate your report. The preliminary pages should be numbered with lowercase roman numerals.

Stages of the report

A good report originates with an appropriate topic. You need to develop a rough outline, conduct a literature review, write a first rough draft, perform editorial cleansing, and develop a final draft.

Progress report

A progress report is a good way to monitor and track milestones in your research. If the report is a progress report, preliminary report, or an intermediate report, the contents should be organized to contain background information, statement of the objective, and specification of the methodology. Within the methodology, be sure to include and update literature review (if any), a revision of the methodology, and a statement of new innovative approaches, as may be applicable. Other elements of the progress report include results to date, work accomplished since the previous report, relevance to the previous work, and continuation from the previous work.

The problems encountered should highlight whether they are with the methodology, research tools, software, hardware, and/or equipment. You should state how the problems were addressed and resolved.

In addition, the proposed work and the expected timeline are important. This will include new ideas investigated, activities involved, and the associated time estimates. You should also specify the expected contribution to your overall research agenda as well as the future research, as applicable.

Use of figures and tables

A picture is worth a thousand words; so goes the saying. We recommend that you use graphical and tabular presentation of information whenever possible. Label all figures and tables. Use concise and descriptive captions for the figures and tables. Each figure or table should be referenced prior to (but close to) its location in the text. When a figure or table is copied from another source, clearly make a reference to the source. You must specify whether the graphics is being replicated as originally published or is an adaptation of the original material.

Use of appendices

Use appendices to elaborate the organization of your report. All incidental or lengthy materials that are not appropriate for the body of the text should be placed in the appendix. These include mathematical derivations, lengthy proofs, computer printouts, flowcharts, and experimental data. The conclusions derived from these materials should be included in the body of the text. Each appendix should have a title and be referenced in the body of the text. Label appendices using uppercase letters, A, B, C or 1, 2, 3, and so on. Appendices should be placed at the end of the report after the list of references.

Use of computer materials

Most science and engineering research nowadays make use of computer tools. You need to use computer tools effectively. Don't throw in computer analysis indiscriminately. Sort outputs printed on computer paper before including in reports. You may scan computer outputs and insert them as graphic images in your report. If possible, odd-shaped computer outputs should be reduced to the standard paper size used for the report (e.g., 8.5 by 11), but should still be readable. It is essential to document the computer code with flowcharts. Use comments within the computer code to document the logic used where applicable. Refer to and observe standard computer-coding practices that increase the readability of the computer code.

Miscellaneous report attributes

The following are miscellaneous items that can enhance your report both in aesthetics and content. The following list summarizes our recommendations for miscellaneous attributes of your report:

- Pay attention to the overall quality of the report in terms of appearance and contents.
- These days, there is hardly an excuse for having spelling errors in your reports. Use the spell check in your word processor.
- Proofread your report before submitting it. Allow yourself enough lead time for this "unpleasant" part of report writing.
- Avoid abbreviations such as "thru" for through, "etc." for and so on, and "plc" for please within sentence structures. Avoid common spelling confusion such as between "principle" and "principal."
- Observe standard guidelines for spacing between sentences and spacing within paragraphs.
- Learn the proper use of comma, colon, and semicolon in sentences.

- Avoid the temptation of using incorrect plurals such as "notations," "informations," "equipments," and "softwares."
- Recognize the proper singular/plural forms of commonly used words: datum/data, criterion/criteria, phenomenon/phenomena, and medium/media. For example, use "data are available" instead of "data is available."
- Be sure to use the proper sentence structure to meet the theme and/ or orientation of your report. For example, for technical journal papers, avoid the use of the first-person singular "I." Use the plural "we," instead.
- Spell out any number that is 10 or less.
- Look over the report to ascertain its appealing structure and first-look appearance.
- Avoid plagiarism at all cost. Never use materials from other sources without proper attribution. Cite your references, credit your sources, and use quotation marks as appropriate.

All summaries, quotations, and ideas borrowed from other sources must be documented. Every source used in the paper should be included in the list of references. Quoted materials should be placed within quotation marks or in an indented block form to clearly identify them as works from other sources.

Writing diagnostics

When students write, they often have problems in several areas. The following are some of the most common problem areas. Coming up with ideas is one common problem. This can be solved by doing a thorough literature search. Reflect on the ideas presented in the literature and try to come up with your own ideas to complement or refute what you read. Finding data to support the research is another common problem. Again, literature search will help here. Prewriting the research paper with hypothetical data will also help you identify what sort of data you need to search for. Next, coming up with a narrow and focused research statement is also a challenge for graduate students. In this case, you need to consult with your research advisor for help. Another problem, yet, is constructing good illustrative examples. For this case, develop enough relevant examples, but avoid giving too many. Recall the logic and premise of the research paper as you find suitable examples. The next problem is organizing the materials into a cohesive sequence of supporting background. For this, start with a broad outline and iteratively rearrange the outline as you expand to more detailed contents. The next problematic challenge for students involves proofreading. Never compromise proofreading. If possible, ask others to read your draft. The more eyes you can place on the draft, the better the quality of the eventual paper. Check the paper for cohesion and coherence of the sections in the paper. Check for spelling, punctuation, and grammatical errors. Do sentences always say what you mean? Is the language appropriate for your audience? Does the format of the paper comply with the guidelines and requirements?

Guidelines for technical review

The research steps usually encompass problem formulation, problem analysis, search for alternate solutions, selection of the best solution methodology, and dissemination (presentation and defense) of the proposed solution. Presentation is the last crucial part of the research effort. Sufficient detail must be provided to fit the needs of the audience being addressed. The higher the hierarchy of the audience, the lower the level of detail should be. Since your research advisor is essentially your research collaborator, he or she must be apprised of the full details of your work. The presentation should follow the general structure of agenda, delivery, and summary. This follows the common adage of informing your audience what you are going to tell them, telling them, and then recapping what you have told them.

Selling your research (figuratively)

Throughout this book, you have been asked to ponder the question, "What is Research?" We suggest that research means inventing or developing a new idea and proving that the idea works or does not work. The "proving" part of this definition of research depends, in part, on how well the idea is presented. That is, how well did you "sell" your research? This assumes, of course, that the idea is worthwhile to begin with. You will identify the focus of your research by being specific and also identify the important elements that support the focus. Describe the nature of the work, if possible, in one sentence. In doing so, you demonstrate mastery of the material to yourself and your audience by distilling your complex work in terms that are relatable and understandable. Explain the methodology briefly and highlight the significance of your research by calling out its contributions to the body of knowledge that you are adding to.

To provide supporting documentation for your research, consider preparing some documentation (e.g., handouts) to corroborate your key points. Give the audience the option of examining specific points in detail later on. Use concise presentation of statements and check spellings throughout the documentation. In preparing this additional documentation, note the documentation may use a smaller typeface than presentation slides.

You need to identify and know your audience. You want the audience to resonate with your research presentation. Reflect on who is attending your presentation (e.g., practitioners, researchers, students, etc.). Know why each person is attending. It would also help to know the background of the audience. Try and figure out what the audience may already know so that you don't waste time rehashing basic materials that are not the "meat" of your research.

You want to grab and retain the attention of your audience. Organize your presentation in a way that draws the attention of your audience to the sequential topics of your research. Use "few words to many ideas" ratio approach. This means that you should avoid pontification. Cover whatever you want to say succinctly and yet informatively. Reiterate key points and elicit reaction or comments from the audience at key points in your presentation.

Participative approach

Get the audience involved. Liven up your presentation charts with color, cartoons, captivating humor, and so on to capture and retain the attention of the audience. This will get the audience involved and entertained while becoming informed. Whenever possible, use the participative approach to convey your message. Use sight, sound, and action to make your message to sink in. Adults absorb, retain, and learn according to the following categorization:

a. 10% of what they read
b. 20% of what they hear
c. 30% of what they read and hear
d. 50% of what they hear and see
e. 70% of what they recite to themselves
f. 90% of what they do. So, get the audience involved if possible.

With respect to Item (f) above, leverage the Chinese axiom that says, "Tell me, and I forget; Show me, and I remember; Involve me, and I understand." The more the audience is engaged and involved in your presentation, the more your message will be understood and retained.

Effective use of time

You need to realize that timing is the essence of an effective presentation. Realize that time is a precious commodity that must be used well. Don't waste the time of your audience. Don't occupy their time needlessly. Both written and oral communication of your work should follow these guidelines. The time of the audience is precious. Let them know that you value their time and their attention. Use presentation time effectively. Be concise and, yet, thorough. Avoid reading from your presentation slides. Rather, talk to the audience. Explain your work with pride and confidence. Long presentations become boring and lose the audience. It shouldn't take long to convince someone of a good idea; if it lasts more than 30 minutes, it's probably too long.

For your presentation materials, use graphical and tabular representations whenever possible. Use an adequately large typeface for slides. Do not include too much on each page (not more than 10 lines per slide). Keep the lines short; no more than seven words if you can help it. Avoid full sentences. For tables, use no more than five columns in a table. Leave adequate space in-between lines and avoid tiny letter sizes for table headings.

Major components of technical communication

Technical presentations vary in depth, detail, and length. You need to be mindful of the needs of your average audience. Technical presentations should progress through the three stages of introduction, development, and results. These three major components can be expanded to cover the general outline suggested below.

- Introduction
  Should get the attention of the audience
  Use a combination of the following:
    Opening gag (humorous or serious)
    Significant results or facts
    Conclusion that demonstrates the importance of your topic
  Establish your credibility
    Background
    Current and/or past-related job functions
    How long the research has been going on
  Identify the focus of the presentation
    State the purpose of the presentation
  Relate your topic to the audience
    State why the topic is important to the field
    State how the audience might use the results
  State the major points (not too many) to be covered
    What the audience can expect to learn
- Body of presentation
  Related works in the area
    Literature search
    Literature review
    Case study

Methodology
>   Existing tools, models, techniques, etc.
>   What is new—background, objective, results to date, proposed work, and contribution
>   Variables of interest
>   Theoretical, analytical, conceptual, actual, experimental, etc.

Results

Validation
>   What is validated?
>   The basis for the validation
>   Lesson for oral presentations: The more lengthy and irrelevant things you talk about during your presentation, the more holes of unanswerable questions you dig for yourself.

- Conclusion
  Review or summarize the main points of the presentation
  Reemphasize the results
  Recommendations
  Identify areas for further research
  Refer back to introductory statements to tie everything together
- Recap
  Be able to say in one sentence what your research objectives are
  Be able to state your research methodology in one sentence
  Be able to state your contribution in one sentence
- Exit
  Departing quip (humorous or serious)
  Leave a lasting impression on the audience

Presentation style

For your presentation style, you should practice whenever possible. This will have the benefits of helping you to reduce nervous tension, helping you to maintain the flow of your presentation, helping you to identify proper transition points, and helping you to stick to time requirements.

Minimize reading from notes or note cards. Maintain eye contact with the audience. Shift your eye contacts around the room so that each segment of the audience would know you are talking directly to them. Don't gaze at an individual too long because it can make the person nervous of being singled out, but don't leave anyone out of your eye contact. Use gesturing (not excessively) and "majestic" movements to get attention. But don't overdo this so that you don't come across as overconfident, complacent, or cocky. You need to maintain an appropriate and consistent vocal volume.

Presentation management

The format presented below is useful for evaluating a presenter's performance at a technical presentation or defense of thesis research. This is useful both for self-assessment and critique of other presenters. The layout can be modified or altered to fit specific evaluation needs.

Excellent: Has a wide and deep knowledge of the subject matter and of the role in relation to his/her special area as well as to the general area of science and engineering. Good knowledge of scientific principles.

Very good: Detailed knowledge of the subject as well as grasp of a broad area. Has a strong awareness of the relevancy of facts to each other and of the underlying principles and implications of results.

Average: Generally accurate information and knowledge of the basic facts of the topic. Some appreciation of the organization of these facts in his/her own area and relevance to other areas.

Fair: Has minimal accurate information. Supporting pieces of evidence or data are weak. Has only occasional evidence of awareness of relevance of knowledge and information to developing plans and utilization.

Poor: Inaccurate and/or irrelevant statements. The presenter presents misinformation. Weak grasp of principles and interrelationships between facts and research.

Grading of oral presentation can be based on the following:

a. Mechanics of presentation (organization, use of visual aids, effective use of time, etc.)
b. Quality of presentation (clarity, handling questions, knowledge of the problem and solution methods, etc.)

Item (b) should carry more weight (e.g., weighted twice as much) than Item (a). It can be evaluated based on the ordinal scale below:

Excellent: Quickly grasps and develops the point of a question. Presents a logical and well-organized response. Is clear, confident, and poised. Uses nice visual aids.

Very good: Understands questions and gives a relevant well-organized response. Appears to be in good command of himself/herself and the situation.

Average: For the most part, presents replies in an organized and suitable way. Needs occasional probing to draw him/her out. Is somewhat self-confident.

Fair: Often responds slowly. Asks the interviewer to repeat and explain questions too often. Appears uncertain of himself/herself, either by being diffident or by obviously bluffing.

Poor: Has difficulty in replying to questions. Misses the point, but does not know it. Needs many probing questions to produce an answer at all. Makes illogical statements. Gives quick, evasive, and diffuse responses.

For a student's presentation, the overall evaluation of performance on an oral exam can fall into one of the categories of high pass, pass, marginal pass, or fail.

You can expect comments and notes on recommended remedial work. Some subtle assessment areas that students don't often consider include the title of the paper or presentation, the date submitted, and the date the submission was signed off. We have seen cases where it was at the presentation event that it was discovered that the research title was erroneously presented. The final evaluation, based on all the evaluative inputs can fall into the category of satisfactory, unsatisfactory, or decision deferred.

Management of group presentations

In cases where technical presentations are to be made by student groups in a classroom setting, the performance evaluation can be based on a peer-rating approach. Each student in the class (the audience) will have an opportunity to evaluate the presentation of every other student in a round-robin presentation format. Every member of each group is expected to contribute effectively to the group project. Each student will submit a confidential evaluation of each member in his or her group at the end of the academic term. The weighted evaluation will then be used in distributing the final group grade to the members of the group.

Communication through publishing

An archival printed publication is the ultimate avenue for disseminating your research as it allows you to reach a much larger audience than you would ever be able to reach in person. Many graduate programs required graduate students, particularly doctoral students, to publish by the time they graduate. Some of the general categories include the following:

- Survey paper
- Research paper
- Experimental paper
- Technical notes

For your publication to be effective, it must appear in an appropriate medium. There are trade magazines and archival journals. Most graduate research results will be targeted for an archival journal publication. The key factors to consider in determining where to publish include the focus and scope of the journal, the circulation volume of the journal, the relevance of the journal's focus to your own area of specialization, the nature and length of the journal's review process, and the lead time to publishing papers accepted for publication. Be sure to get a copy of the author's guide for the journal where you expect to publish. Follow the guidelines as much as possible. For cases where a call for papers is issued by the journal, try and operate within the specified guidelines and deadlines.

You need to take advantage of software tools whenever such tools are available. Commercial software packages can add flexibility and functionality to your document-processing needs. A complete research facility should have the different types of applications software from basic word-processing tools to complex graphics tools. You should use them in such a way that they complement each other. You can use any of the common tools to draw charts and figures for your research presentations. The graphics files can be merged with word-processing documents to produce complete publishable research reports. Spreadsheets can be used to format numeric tables of experimental data. These, like graphics, can be merged with reports. Sometimes, researchers have database files that they would like to include in their word-processing documents to create integrated reports. Such database files can be retrieved with a database manager program, organized into an appropriate format, and then exported to the word- processing document.

If you would like to have all your application programs right at your fingertips, so to speak, you may want to procure an integrated application software package. Such a single program may include modules for several application programs such as word processing, e-mail communication, database management, spreadsheets, presentation, and graphics. An integrated package permits easy manipulation and interchange of data in several different applications. Moreover, they allow the user to learn one consistent set of commands for accomplishing a variety of tasks instead of learning the individual sets of commands for independent packages. But, integrated programs can become cumbersome to install and use, particularly if you need only one of the several functions of the package. For example, MATLAB® is a robust software package that provides several functionalities for science, technology, engineering, and mathematics (STEM) research.

Managing poster presentations

Poster presentations are often used as a quick avenue to communicate research work at technical conferences or professional meetings. A poster should concisely convey the objectives, methodology, data, results, and conclusions of a research effort. A poster can serve as your surrogate presenter. Whether the poster is tended or not tended, it should be self-explanatory. It should explain your research to the audience. Most likely, your poster audience will be single individuals visiting the poster and following the contents on their own. Even if you are there by your poster, you will notice that some visitors prefer to read through the poster (uninterrupted) on their own. A poster presentation is a condensed technical paper presented through a variety of graphic wall or table displays. These may include graphics, charts, photographs, computer outputs, scale models, and samples. The displays make the presentation of certain types of information more effective, such as mathematical representations, graphical models, large data sets, or lengthy materials. This allows the audience to review the poster displays at their own pace and to discuss the materials with the authors in detail on a one-to-one basis. A poster session provides a unique opportunity for authors to present a technical paper, while affording the attenders an easier format for asking questions and receiving in-depth responses. Many presenters and attenders feel more comfortable in a self-paced discussion than a condensed formal presentation. We present a complete guide for managing poster presentations in Appendix 12.

Importance of publishing your work

Documenting and publishing your work are as important as doing the work in the first place. Those in academia know this axiom very well; so much so that a dreaded truism in academia is "Publish or Perish." Getting a graduate degree often comes with a requirement or an expectation to publish. Whether you plan to end up in academia or not in your future career plans, we advise you to commit yourself to publishing your work at some point. Publication not only disseminates your work, but also serves as an archival testament of the initial origin of the work. To mitigate and "positivize" the dreaded saying of "Publish or Perish," Badiru (2008) suggested the alternative of "Publish and Flourish." This is to emphasize that good things come from publishing your work, even if you are not in academia. Publishing your work secures your intellectual property and ownership of the work in case infringements come up in the future. In academia, the "flourish" part is evidenced by the career advancement that comes with journal publications. In nonacademic professions, the "flourish" part can come if your work, later on, becomes the basis for some new and marketable commercial product. In essence, whatever your career path, you should plan to publish your graduate work as a pathway to flourishing professionally.

Conclusion

Communication is the forward wheel of every research project. The task of communicating your work should be handled as a project in itself, with all the requirements of planning, organizing, scheduling, controlling, and terminating. Projects are executed and accomplished through the collective efforts of people, tools, and processes. You, as the graduate researcher, fall within the "people" group of this three-pronged spectrum. You must use the available tools and processes to make your research communication more effective. Communication is the glue that binds all your research efforts together. Various reports, anecdotal case examples, and formal studies have indicated that at least 90% of project failures can be traced to poor communication. Even after your research is completed, it is not done until it is communicated effectively. Be sure that your work is communicated with respect to the elements of who, what, why, when, where, and how. The foundation for the success of your research is proper communication, leading to a sustainable appreciation of your work. As presented in this chapter, it is important to not only do a good job of communicating your work, but also to understand how the audience perceives and rates your work. For this reason, this chapter presents evaluation and assessment criteria typically used by the audience to assess what is being communicated.

Reference

Badiru, A.B., Publish and flourish: A new paradigm for OR/MS faculty. *OR/MS Today*, Vol. 35, No. 1, February 2008, 18–19.

Appendix 1: 50 ways to improve your research project

1. Abide by the rules of the project
2. Align corporate projects with organizational needs
3. Align home projects with family needs
4. Allocate sufficient resources to meet project objectives
5. Be good, so that you can receive goodness from others
6. Be specific with requirements
7. Check on project milestones
8. Commit to whatever you are doing
9. Communicate to solicit cooperation
10. Cooperate so that you may receive cooperation in return
11. Coordinate with others so that the load can be shared
12. Define goals clearly upfront
13. Delegate so that others may learn the path to success
14. Do not cut corners; corners can come back around
15. Document the project for future reference
16. Don't despair; there is success at the end of the project tunnel
17. Embrace ethics as a platform for project success
18. Embrace new ideas
19. Enjoy leisure as a break from project monotony
20. Evaluate the consequences of project actions
21. Focus on the end goal
22. Get organized
23. Homestead for personal projects
24. Integrate project outcomes with the operating environment
25. Justify each time or resource expenditure on the project
26. Keep project scope within reason
27. Know that time is everything; once lost, it cannot be recovered
28. Make accountability a requirement for project execution
29. Manage yourself as a critical resource for your project
30. Measure everything to provide a metric of assessment

31. Network with potential project allies
32. Operate lean and cut-out fluff
33. Place tools where they belong
34. Plan and plan again
35. Practice continuous improvement
36. Preempt project problems by asking "what-if" questions
37. Promise only what you can deliver
38. Question everything with a constructive open mind
39. Recognize that there is always room for improvement
40. Replan when the plan is not going well
41. Schedule each activity so that it can get done
42. Set standards so that your project will have a target
43. Sleep enough so that you can be rejuvenated for your project
44. Take care of your health as a project asset; no health no success
45. Take corrective actions at the earliest opportunity
46. Take responsibility for what you are responsible for
47. Treat your people well; they are your most enduring project resource
48. View everything as a project
49. Observe safety requirements; accidents divert attention
50. Simplify your process.

Appendix 2: How to get topic approval

To get approval of your research advisor for your proposed research topic, develop a topic justification outline similar to the one presented below. Make it concise enough to fit within not more than two pages. This outline is not a formal research proposal, but a brief indication of the justification for the research.

Topic: Specify the topic area or title

Problem area: Identify the problem area to focus on

Importance of the problem: Describe the implications of the problem

Existing or conventional approach: Describe the method of solving the problem

Shortcomings of the existing approach: Explain the drawbacks of the existing method proposed

Methodology: Outline your proposed approach

Validation: Describe your validation strategy if applicable

Expected results: Identify what is expected from the research

Criterion measure: Identify the performance measures for the research

Contribution: Establish the contribution of the research

Appendix 3: Research proposal evaluation checklist

The research proposal must be evaluated the same way that any project proposal is evaluated. On a scale of 1–5 (1 = poorest, 5 = best), rate each component listed in Table A3.1 for the various elements of the research proposal. Not all the elements included below will be applicable to all proposals. It may be necessary to condense or expand the list based on specific situations and interests.

Table A3.1 Template for research proposal evaluation checklist

	1 (Poor)	2	3	4	5 (Best)
Abstract or research summary					
1. Is preceded by research title					
2. Appears at the beginning of the proposal					
3. Identifies the research area					
4. Includes a problem statement					
5. Includes a statement of the objective					
6. Includes a statement of methodology					
7. Is clear and concise					
8. Is relevant to the area of specialization					
Statement of the problem	1 (Poor)	2	3	4	5 (Best)
1. Concisely states what is to be done					
2. Is of reasonable scope					
3. Is supported by literature review					
4. Indicates a new idea, creativity, or innovation					
5. Gives evidence of knowledge of the field					
6. Is relevant, interesting, and timely					
7. Is feasible and shows potential for success					
8. Makes no unfounded assumptions					

(*Continued*)

Table A3.1 (Continued) Template for research proposal evaluation checklist

Research objective	1 (Poor)	2	3	4	5 (Best)
1. Relates to the statement of the problem					
2. Clearly connects to the expected outcomes					
3. Distinguishes objectives from methods					
4. Describes expected benefits					
5. Identifies the performance measure					
Research methodology	1 (Poor)	2	3	4	5 (Best)
1. Connects with the problem statement and objective					
2. Describes the experimental or research protocol					
3. Contains evidence of scientific principles					
4. Describes justification of the approach					
5. Describes the data management plan					
6. Describes research performance facilities					
7. Includes a project management plan					

Appendix 4: Benefits of industry-sponsored research

Graduate research projects sponsored by a business or industry entity deserve special attention. The company must be convinced of the importance and benefit of the research. The presentation of such research results must focus on their impact on the company's bottom line. Industry-sponsored projects can cover three primary objectives:

1. To identify current methods, approaches, and organizational linkages used to implement manufacturing and/or industrial extension education programs in universities.
2. To assess the universities' current capacity, available resources and expertise, and willingness/ability to meet future requirements for an effective industrial extension program.
3. To identify appropriate intrauniversity and intercompany linkages vital to the implementation of university–industry collaborative research.

For industry-sponsored projects, you must provide answers to the following questions.

About the problem

- What exactly is the problem?
- How important is the problem to company operations?
- What specific evidence of the problem is available?
- What has already been done about the problem?
- What happens if the problem is not addressed?

About the methodology

- What are the expected benefits of the proposed solution?
- What are the anticipated difficulties, limitations, and requirements?

- What are the risks associated with the solution proposal?
- What contingency plans can be made for the research?
- How is the proposed methodology justified over other approaches?

About the implementation

- When can the proposed solution be implemented?
- How long will the implementation take?
- What resources, costs, and actions are needed for the implementation?
- What are the success metrics (evaluation approach) for the implementation?
- What are the milestones to measure progress of the implementation?

About the benefits

- What are the short- and long-term impacts of the solution?
- What is the return on investment for the solution?
- What is the benefit–cost ratio for the solution?
- What guarantees (if any) are available for the expected returns?
- What are the benefits for future research collaborations?
- Who are the beneficiaries of the project?

In dealing with industry personnel, it is quite possible that different levels of management will be encountered. Presentations should be tailored to the interests of the different groups and management levels. A good rule of thumb is that the higher the level of management, the lower the level of detail for research presentations. Table A4.1 presents items to focus on selected groups of industry personnel.

Table A4.1 Suggested contents of research presentation to industry

Senior management	Middle management	Technical personnel
Return on investment	Increased production	R&D benefit
Alignment with a strategic plan	Reduced personnel problems	New technology adoption
Increase in market share	Better customer service	Technical detail of the approach
Improved decision tools	Employee satisfaction	Project schedule
Long-term growth	Cost control	Computer interface
Capital improvement	Better operational planning	Training requirements
Cost reduction	Better production schedules	Research documentation
Community relations	Better information access	Alignment with the process

Academic capability for industrial collaboration

To assess the current capability and existing resources for industrial collaboration, consider the questions presented in the next section. The expected contribution of a university in industrial collaboration is to help a company improve its products, services, processes, and, consequently, profits. Bilaterally, the expected contribution of a company is to provide funding and a practical test bed for university research. In the industry, the idea of zero defects makes sense. But in academia, zero defects make no sense since we cannot guarantee the research success of each student. This notion can affect the expectation of the industry concerning university expertise and products.

Characteristic keywords for industry

- Profit driven
- Competition conscious
- Cost awareness
- Practicality focus

Characteristic keywords for academia

- Knowledge oriented
- Exploration driven
- Constrained time environment (academic year cycle)
- Individualistic research focus

From a collaborative project management viewpoint, answers to these questions will shed light on the respective capabilities of the two seemingly incompatible research collaborators.

1. What specific industrial problems will the joint effort focus on? What is the cause of the problem? For example, regulatory requirement, new manufacturing process, infrastructural development, use of new technology, or plant, equipment, and training needs may call for a company-sponsored research.
2. How is the need for the collaborative research identified? Did the company approach the university? Did the university solicit the project?
3. What is the stimulus for university involvement? Is it initiated by an individual faculty member, an organized program (e.g., a research center), or a need to find an application for a research project?
4. What can actually be done in the collaborative project? Examples might include a combination of specific technology, process, system applications, management consultancy, worker training, education

program, plant layout, cost analysis, or evaluation of alternative technologies.

5. How will the university organize to address the industry problem? Is there an existing organizational structure (e.g., a laboratory) within which the project can be carried out? Is there an individual or faculty team to manage the interface between the university and the industry? Is there an interdisciplinary arrangement to tap into the university expertise from relevant fields?

6. What university resources, expertise, departments, or disciplines will be involved in the university–industry collaborative effort? Will the university be reimbursed for any of the associated costs of using these resources?

7. What external resources and expertise will be needed for the collaborative project? How will the necessary linkages be created and maintained?

8. What are the expected impacts of the project (e.g., jobs, profits, new products, better quality, cost reduction, etc.)? What training opportunities will be available to students as a result of this project?

9. How long is the project expected to last? Can university resources and company resources be tied up for that long?

10. Does the university have the interest and willingness to commit additional resources and initiate organizational requirements to pursue future industrial collaboration?

Appendix 5: Sample three-semester master's thesis schedule

Deliverable	Description	Due by Friday of week
Semester 1		
Thesis prospectus	1-page memo containing: draft thesis title, advisor, description of research plan, sponsor, required resources/hardware/software/labs/travel/processes (IRB approval)	1
Outline Chapter 1	Detailed outline containing subheadings, and main topics for each subheading	2
Literature survey proposal	1-page table of specific literature topics you will search for, how they relate to your topic/what you expect to learn, alternative search terms, and proposed databases	2
Literature inventory 10–20 sources	List/spreadsheet of relevant papers. Inventory should include bibliography info and 1–2 sentences on why/how paper is relevant	3
Outline Chapter 2	Detailed outline containing subheadings, and main topics for each subheading. Indicate which topics you already have references for and which still need references	4

(*Continued*)

Deliverable	Description	Due by Friday of week
Draft Chapter 1	Complete first draft, including feedback from outline in the required thesis format	4
Updated literature inventory	Continue to build literature inventory, using feedback from Chapter 2 outline	6
Draft Chapter 2	Complete first draft, including feedback from outline in the required thesis format	8
Outline Chapter 3	Detailed outline containing methodology subheadings, and main topics for each subheading. Methodology should include process for collecting and analyzing data for each investigative question/phase	9
Draft proposal presentation	Presentation should include background/motivation, literature review, research gap, research question and investigative questions, methodology	10
Semester 2		
Proposal presentation to thesis committee	Presentation should include background/motivation, literature review, research gap, research question and investigative questions, methodology	1
Draft Chapter 3	Complete first draft, including feedback from outline in the required thesis format	3
Methodology development	Creation of models, algorithms	2–3
Methodology execution	Conduct experiments, run models	3–5
Analyze data	Perform statistical analysis	3–6
Outline Chapter 4	Detailed outline containing subheadings, and main topics for each subheading	6
Draft Chapter 4	Complete first draft, including feedback from outline in the required thesis format	8
Revised Chapter 3	Including feedback from first draft in the required thesis format	9

(Continued)

Deliverable	Description	Due by Friday of week
Revised Chapter 4	Including feedback from first draft in the required thesis format	10
	Semester 3	
Update deadlines	Obtain deadlines from thesis processing center	1
Outline Chapter 5	Detailed outline containing subheadings and main topics for each subheading	2
Draft Chapter 5	Complete first draft, including feedback, from outline in the required thesis format	4
Revised full five-chapter thesis	Complete first draft, including feedback, in the required thesis format	5
Final draft thesis to committee readers	Revised final draft	6
Graduation application	Submit graduation application	TBD
Draft thesis presentation	Update proposal presentation	7
Thesis defense	Oral defense	TBD
Submit electronic documents	Submit to thesis processing center	TBD

Appendix 6: Sample work breakdown structure

<table>
<tr><td></td><td>Start date</td><td>End date</td></tr>
</table>

1. Classes
- 1.1 First quarter classes
    - 1.1.1 Register for classes
    - 1.1.2 Introduction to statistics
        - 1.1.2.1 Lectures
        - 1.1.2.2 Homework
        - 1.1.2.3 Exams
        - 1.1.2.4 Term paper
    - 1.1.3 Class #2
        - 1.1.3.1 Lectures
        - 1.1.3.2 Homework
        - 1.1.3.3 Group project
        - 1.1.3.4 ...
- 1.2 Second quarter classes
    - 1.2.1 Register for classes
    - 1.2.2 ...
    ⋮
- 1.6 Sixth quarter classes
    - 1.6.1 ...
    - 1.6.2 ...
    - 1.6.3 ...

2. Research
- 2.1 Introduction
    - 2.1.1 Define research problem
        - 2.1.1.1 Determine problem context of problem
        - 2.1.1.2 Initial literature review of proposed problem
        - 2.1.1.3 Get topic approval from advisor

2.1.2 Research questions
    2.1.2.1 Develop main research question
    2.1.2.2 Develop supporting research questions
    2.1.2.3 Get research question approval from advisor
2.1.3 Writing
    2.1.3.1 Draft introductory chapter
    2.1.3.2 Review introductory chapter with advisor
    2.1.3.3 Distribute chapter to committee members

2.2 Literature review
  2.2.1 Understand literature search resources
    2.2.1.1 Meeting with librarian
    2.2.1.2 Attend library seminar
    2.2.1.3 Find off-campus resources for literature
  2.2.2 Searching and reading
    2.2.2.1 Topic A
    2.2.2.2 Topic B
    2.2.2.3 Topic C
    2.2.2.4 Topic D
  2.2.3 Integration and notetaking
    2.2.3.1 Develop personalized notetaker
    2.2.3.2 Create and populate concept matrix
  2.2.4 Writing
    2.2.4.1 Draft literature review chapter
    2.2.4.2 Review literature review with advisor

2.3 Methodology
  2.3.1 Develop methodology
    2.3.1.1 Literature review on potential methods
    2.3.1.2 Discuss potential methods with advisor
    2.3.1.3 Down select to single method for study
    2.3.1.4 Approval from advisor on method
  2.3.2 Writing
    2.3.2.1 Draft methodology chapter
    2.3.2.1 Review chapter with advisor

2.4 Data collection
  2.4.1 Conduct data collection
  2.4.2 Writing
    2.4.2.1 Describe data collection process
    2.4.2.2 Review chapter with advisor

2.5 Analysis
  2.5.1 Conduct analysis
  2.5.2 Writing
    2.5.2.1 Describe data analysis process
    2.5.2.2 Review chapter with advisor

2.6 Results and conclusions
   2.6.1 Writing
      2.6.1.1 Draft results section
      2.6.1.2 Draft discussion section
      2.6.1.3 Draft conclusions section
      2.6.1.4 Draft future work section
      2.6.1.5 Draft recommendations section
   2.6.2 Review chapter with advisor
2.7 Committee meetings
   2.7.1 Discuss with advisor who should form committee
      2.7.1.1 Schedule one-on-one meetings to discuss topic
      2.7.1.2 Send invitation to potential members
      2.7.1.3 Confirm final makeup with advisor
   2.7.2 First committee meeting
      2.7.2.1 Develop slides/agenda for first committee meeting
      2.7.2.2 Dry run with advisor
      2.7.2.3 Schedule meeting
      2.7.2.4 Send read-aheads (slides, draft chapters, agenda)
      2.7.2.5 Hold meeting
   2.7.3 Thesis defense
      2.7.3.1 Submit draft thesis to committee
      2.7.3.2 Develop slides for thesis defense
      2.7.3.3 Schedule thesis defense
      2.7.3.4 Send read-aheads (slides, revised thesis)
      2.7.3.5 Hold thesis defense
   2.7.4 Graduation
      2.7.4.1 Determine schedule for graduation deadlines
      2.7.4.2 Apply for graduation
      2.7.4.3 Final signatures/approvals on thesis document
      2.7.4.4 Final drafts to thesis processing office
      2.7.4.5 Graduation ceremony

Appendix 7: Sample thesis outline

1. Chapter 1: Introduction
   1.1. Background
   1.2. Description of research problem
   1.3. Description of research gap and research objectives
   1.4. Research question and subquestions
   1.5. Short description of methodology (e.g., human trial experiments, simulation, case study, regression analysis on existing data set, and interviews/surveys)
   1.6. Assumptions/limitations
   1.7. Expected contributions/study results
   1.8. Organization of remainder of document
2. Chapter 2: Literature review
   2.1. Introduction
   2.2. Topic A
   2.3. Topic B
   2.4. Topic C
   2.5. Description of gap
   2.6. Summary/conclusion
3. Chapter 3: Methodology
   3.1. Introduction
   3.2. Overview of research methodology—suggest diagram, see example
   3.3. Overview of system/case study
   3.4. Description of dependent and independent variables
   3.5. Experimental design/description of data set and source
       3.5.1. Experimental tasks
       3.5.2. Experimental equipment
       3.5.3. Experimental procedures
   3.6. Assumptions
   3.7. Description of how to perform analyses (statistical methods, etc.)
   3.8. Summary/conclusion

Literature review topics could be what current academic literature says about:

- Defining key terms
- Relationship between IVs and DVs
- Methodological approaches

Appendix 8: Tips for literature review

Importance of literature review

The product of a research effort is the expansion of the current body of knowledge. Research is not merely for the purpose of data collection and information generation. Far from it, research is for the purpose of answering a question that has not been answered before. Thus, a key aspect of research is that it requires a unique or new question. Therein lies the need to conduct a literature research to ascertain what has been done or not done regarding the question being posed. You must perform an analysis on the data and information collected to answer the question to some conclusive level. A literature review is essential to get a research pursuit started in the right direction.

When a literature search is done about a subject, two things can happen. You will either find very limited published materials on the subject or you will encounter an overwhelmingly large number of relevant materials. You must learn to prune the literature and sort out the items that are really useful to your research. Once you narrow down the materials, you should briefly summarize each one. This is useful for cross-checking methodologies later on and in writing the literature review section of the technical report. The example format presented below is useful for that purpose in the case of a seminal research on the topic of search techniques for artificial intelligence application.

Template for literature review report

Publication Title: "A New Computational Search Technique for AI Based on Cantor Set"
Author: Adedeji B. Badiru
Source Journal: *Applied Mathematics and Computation,* Vol. 57, 1993, pp. 255–274.

Synopsis: This paper discusses the development of a new search algorithm for artificial intelligence systems. The search technique is based on the mathematical theory of Cantor set. The search algorithm will be efficient for specialized search domains where the distribution of the data elements to be searched is approximately normal. The approach uses the iterative procedure of deleted middle thirds. This facilitates quick pruning of a search space. The new search technique has potential applications in computational systems with large input–output data handling. Extensive experiments comparing the Cantor set search (CSS) to binary search indicate that the search technique holds good promise.

Solution Approach: The paper uses a methodology based on the theory of Cantor set. Search distribution is limited to bell-shaped curves.

Computational Experience: Comparison to binary search.

Conclusions: The new search technique holds promise based on comparative results with binary search. Further research and extension to other types of distributions is needed. Computational complexity of the Cantor set search algorithm is yet to be developed.

Structuring the research proposal

Based on the findings of the literature review, you must present an articulate communication of what is proposed. A research proposal is very much like a conventional project proposal in terms of the general structure and management plan. Even the best idea is worthless if it cannot be sold to anyone. If a research proposal is done well, the job of writing the research report later on will be simplified. For most research efforts, it is effective to document results as the research progresses. In the proposal, you should do the following:

- Develop a narrow, focused thesis statement.
- Be able to say in one sentence what your research objectives are.
- Be able to say in one sentence what your research methodology is.
- Be able to say in one sentence what your contribution is.

If you have difficulty in accomplishing the above, you need to seriously rethink your research agenda. A basic requirement of a successful research pursuit is being able to concisely state your research and convince the audience that it is a worthwhile effort. In addition to definitional discussions and narrative on the problem area, you need a rigorous section on literature surveys to demonstrate familiarity with other researchers' works. The outline of your research proposal may look something like the following:

Part I: Introductory Section
- Introduction
- Background
- Problem statement
- Objectives
- Merit or utility of the research

Part II: Methodology Section
- Proposed methodology
- Literature survey
- Relevance to previous works
- Unique aspects of the research
- What is new?

Part III: Research Value Section
- Expected contribution
- Implementation requirements
- Limitations (if any)

Part IV: Summary Section
- Conclusions
- Recommendations
- Future research

Part V: References

Below are some possible questions to be prepared for in defending your research plan:

- What is new in your research?
- What contribution will your research make to the literature?
- Where and how can your research results be implemented?

Appendix 9: Research methodologies and strategies

> Everything should be made as simple as possible,
> not one bit simpler.
>
> **Albert Einstein**

For any type of research, you need to develop good methodologies and strategies. Write a concise objective statement and review it frequently (e.g., once a week) to avoid irrelevant digression from the focus of the research. Table A9.1 presents an outline for research performance.

The overall layout out of your research methodology might look similar to the outline below:

- Abstract
- Introduction
- Methodology research protocol
- Experimental procedure
- Results
- Conclusion

The key questions for a prospective research are:

1. What do you intend to do?
2. Why is the work important?
3. What has already been done?
4. How are you going to do the work?

To address these questions, organize your research strategy as follows:

1. Provide a clear problem statement
   a. Start with an open mind
   b. Don't jump to conclusions

Table A9.1 An outline for a research methodology

Research element	Description/explanation
State the research hypothesis	Specify the research question to be addressed (what, why)
Define the problem scenario	Explain the situation that is relevant to the problem (who, what, where)
State the research goal	Specify what is to be done (what)
Outline the research plan	Develop the plan to carry out the research (who, what, when, where, how)
Specify the conclusion	Draw conclusions from the research result
Document the research	Report, disseminate, and communicate the research

    c. Identify the problem to be solved
    d. Develop your hypotheses
    e. What you want to do
    f. What you can accomplish
2. State the significance of the proposed work
    a. Problem background
    b. Relevant literature
    c. Gaps that need to be filled
3. Provide a justification for the problem
    a. The discipline and scope
    b. Fields relevant to the research
    c. Intrinsic merit
4. Perform a feasibility of the proposed research
    a. Validity of approach
    b. Qualification and preparation for the research
    c. Available resources
    d. Preliminary study
5. Develop an experimental plan
    a. Project organization
    b. Innovative aspect of the methodology
    c. Feasibility, adequacy, and appropriateness of the approach
    d. Difficulties anticipated
    e. Contingency approaches
    f. Timeline (activities and schedule)
6. Conduct a research evaluation
    a. Data analysis
    b. Interpretation of results
    c. Milestones

7. Do a follow-up on the research
   a. Review the obvious
   b. Continuation of research
   c. Long-range impact
   d. Dissemination

With respect to managing the research enterprise, Figure A9.1 illustrates the relationships between problem, data, and knowledge in science and engineering graduate research. To have a feasible research goal, a workable intersection must exist between the available data, the existing knowledge, and the problem definition. Within the global problem structure, a specific focused research problem must be identified. In the research work, it may be necessary to identify supporting areas that lend data and knowledge to the research at hand.

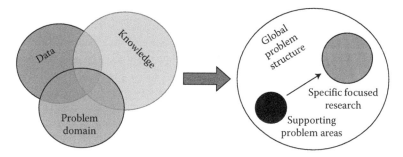

Figure A9.1 Problem, data, and knowledge relationships.

Appendix 10: Sample methodology phasing

An example of how a research methodology can be divided into separate phases is depicted in this appendix. Note that each phase contains inputs and outputs. Phases that result in answering investigative questions/sub-questions are also indicated. Appendix 14 provides a sample section of a research proposal presentation that includes the content described in this appendix.

Proposed methodology

This study is divided into five phases:

1. Baseline model creation
2. Baseline model validation
3. Baseline model workload evaluation
4. Adaptive automation experiments
5. Analysis

Figure A10.1 provides a graphical representation of these five phases; further descriptions of each phase are provided below.

Phase 1: Baseline model creation

Overview: The first phase, baseline model creation, includes establishing the conceptual model of the specific ISR tasks and building a preliminary discrete event simulation based on this conceptual model. The ISR tasks comprise of virtually simulated UGV change detection and threat detection tasks performed by human subject participants as part of an ongoing research project performed by UCF IST ACTIVE Lab studies for the Army Research Laboratory.

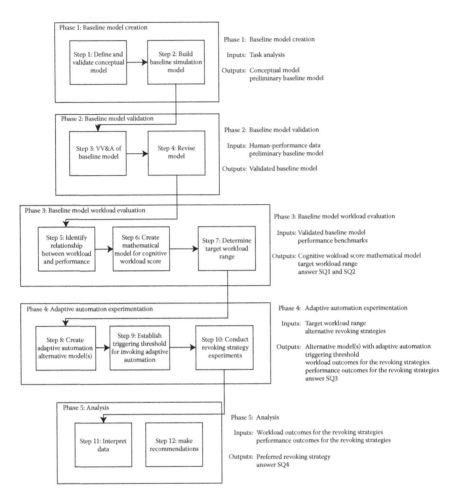

Figure A10.1 Methodology flowchart.

Inputs: This phase requires a detailed *task analysis* capturing subtasks, events, event triggers, user responses, subtask sequencing, cognitive resources engaged, and estimated process times.

Outputs: The outputs of this phase include a *conceptual model* of the change detection and threat detection scenarios that has been validated by UCF IST ACTIVE Lab subject matter experts, as well as a *preliminary baseline model*. The preliminary baseline model will include all subtasks identified during the task analysis and will provide performance measures based on input data. The performance measures include task accuracy for each subtask and cognitive workload level profile over the duration of the replication (workload values by resource channel for each event in the simulation event log).

Phase 2: Baseline model validation

Overview: This phase entails revising and validating the preliminary baseline model based on human-performance and cognitive workload data for the ISR tasks.

Inputs: Validation of the *preliminary baseline model* will be done through a comparison of the model outputs (task accuracy and workload) to the *human-performance data* collected by the UCF IST ACTIVE Lab. The human-performance data includes a number of continuous physiological measures, including brain activity, heart rate, blood oxygenation in the brain, and eye activity. The ACTIVE Lab is currently in the process of collecting this physiological data, which will be compared with subjective workload data in order to identify the physiological measure(s) that are most sensitive to changes in workload. The preliminary baseline model validation will utilize the most promising measure(s). In order for the preliminary model to be considered valid, the model outputs must follow similar directional changes and magnitudes as the human-performance data. The model will be revised as necessary, to reflect real-world observations.

Outputs: This phase concludes with a satisfactory, *validated baseline model*.

Phase 3: Baseline model workload evaluation

Overview: The purpose of this phase is to understand the relationship between the simulation values of workload and performance in order to identify a desired workload range that achieves peak performance. The simulation will identify workload values for each resource channel (visual, auditory, cognitive, psychomotor); thus, this phase will also include establishing a single measure that captures the appropriate relationship/weights of these channels on performance.

Inputs: To perform the workload evaluation, this phase requires a *validated baseline model* providing workload and task performance data, with which to conduct correlational analyses. Establishing a target workload range will also require *performance benchmarks/standards* for the specific ISR tasks. These benchmarks will be generated from the ACTIVE Lab's human-performance data.

Outputs: This phase produces a *cognitive workload score mathematical model* that accurately combines the individual resource channel workload scores into a single measure. The phase concludes with an established *target workload range* based on cognitive workload score, which produces peak task performance. This phase will *answer the first two research subquestions*.

Phase 4: Adaptive automation experimentation

Overview: The purpose of this phase is to incorporate adaptive automation into the baseline model, and to evaluate adaptive automation revoking strategies.

Inputs: Invoking the adaptive automation will be based on the *target workload range*. However, it will be likely that there are delays in the effects of automation on workload and performance, thus requiring a triggering threshold that differs from the target workload range threshold. This phase also requires a detailed definition of the *alternative revoking strategies*.

Outputs: This phase will establish an alternative model with adaptive automation, including a triggering threshold for invoking automation. The main product of this phase is workload and performance outcomes for the revoking strategies based on time, task completion, and workload score. This phase will answer the third research subquestion.

Phase 5: Analysis

Overview: The purpose of this phase is to analyze and interpret the data produced in the revoking strategies experiments in order to identify a preferred revoking strategy.

Inputs: The *workload and performance outcomes for the revoking strategies* will be analyzed using statistical methods to identify which, if any, revoking strategies produce a statistically significant difference in workload and task performance. Appropriate statistical methods will be chosen based on the properties of the data. Average workload, maximum workload, minimum workload, total time in target workload range, and task accuracy will be examined. The primary performance metric is the total time in target workload range.

Outputs: Completion of this phase will produce a recommendation for the *preferred revoking strategy* based on statistical analyses. This phase will *answer the fourth research subquestion*.

Appendix 11: Sample methodology section of research proposal presentation

This appendix shows how to take the content created in Appendix 10 and incorporate it into the Methodology section of your research proposal presentation (Figures A11.1 through A11.7).

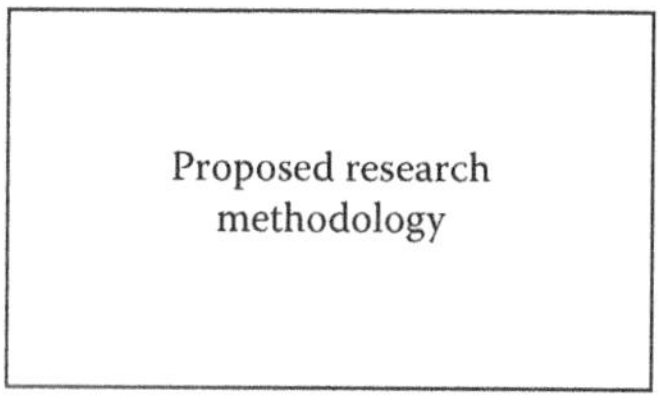

Proposed research
methodology

Figure A11.1 Title slide.

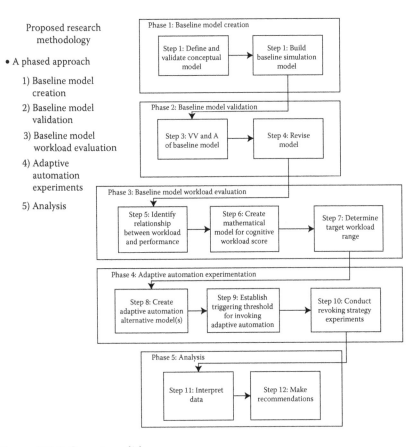

Figure A11.2 Overview slide.

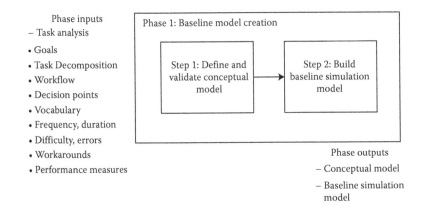

Figure A11.3 Proposed research methodology, Phase 1.

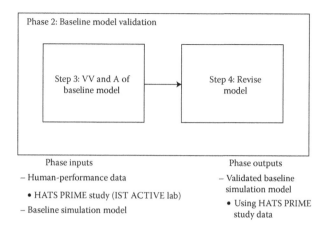

Figure A11.4 Proposed research methodology, Phase 2.

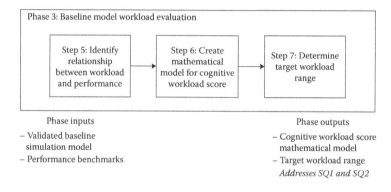

Figure A11.5 Proposed research methodology, Phase 3.

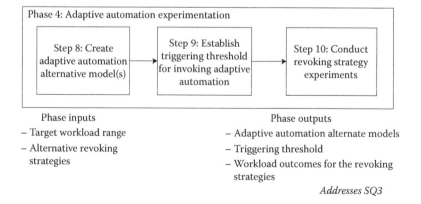

Figure A11.6 Proposed research methodology, Phase 4.

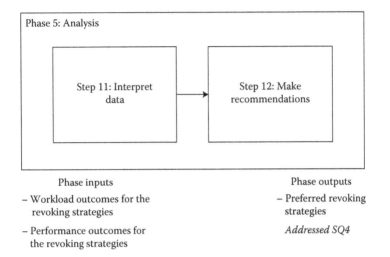

Phase inputs

– Workload outcomes for the
  revoking strategies

– Performance outcomes for
  the revoking strategies

Phase outputs

– Preferred revoking
  strategies

Addressed SQ4

Figure A11.7 Proposed research methodology, Phase 5.

Appendix 12: Guidelines for creating an academic poster

What is an academic poster?

An academic poster is a graphically oriented summary of a research project. It consists of a collection of frames containing brief descriptions and meaningful images and graphics. It is typically presented at a poster session.

Poster sessions are often part of academic or professional conferences. They typically take place in a room that is laid out tradeshow-style, with all posters displayed simultaneously. Interested parties are free to come and go and presenters may spend part, but typically not all, of the time alongside their respective poster.

How are academic posters different from PowerPoint presentations?

Unlike PowerPoint presentations, which display only select piece of information at a time, in a sequenced order, an academic poster displays all information at once. Whereas PowerPoint presentations present information in keyword-bullets, academic posters contain text in a narrative, sentence-paragraph format. Because academic posters are meant to produce quick understanding, a greater emphasis is placed on visual information, including images, graphs, figures, and tables.

As to the presentation itself, the presenter will repeat the presentation many times, as various individuals approach their poster. It is a good idea to get to know your audience; thus, you should begin with a two-way greeting or introduction, if possible. We recommend that poster presenters have a 2–5-minute "elevator speech." Based on your two-way greeting, you can tailor this elevator speech to that particular audience. If the audience shows interest in more detail, be prepared with a more in-depth 10–15-minute explanation of the poster. This explanation should lead the audience logically through the content of the poster, using the visual aids on the poster as guides. If you have a representative image/figure/graph/ table for each section of your poster, you should have no problem leading this guided narrative without the need to read any of your text. You should also be prepared to answer questions and have business cards available to hand out.

How do I create an academic poster?

There are a number of poster templates available (most are in a PowerPoint format) on the Internet to get you started; just use the keyword search "academic poster template." Note that posters can vary in size from 18" × 24" to 36" × 48", so be sure to select an appropriate-sized template for your venue. There are a number of quality resources for information on poster design as well. We recommend this resource from Penn State University for detailed information on organization, layout, font selection, color, and use of images: http://www.personal.psu.edu/drs18/postershow/.

The poster will consist of a poster title, headings, body text, diagrams, charts, figures, etc. These should be sized and lettered so they are legible and readable by a group of attendees at a distance of about 5 feet. They should also be simple, colorful (if possible), well labeled, neat, and well organized. Remember, the poster should succinctly highlight: Problem, Objective, Methodology, Data Analysis, Results, and Conclusions.

Do I need to bring anything to the poster preparation?

Of course, you should bring your poster. In addition, it is a good idea to verify with the poster session organizer how the posters will be displayed and what materials will be provided. If easels are provided, you may need a hard backing to attach your poster to, along with tacks, clips, tape, or Velcro. If wall space is provided, confirm that the means of attaching the poster is also provided, or verify what type of adhesive device is appropriate.

It is a good idea to have handouts to supplement the poster displays. Posting of the entire paper on the poster board is not recommended. Members of the audience always appreciate receiving copies of handouts. A handout gives an attendee something to look forward to (in terms of reading) about your research. The impression of your research will probably last longer with someone who has received a copy of your handout than with someone who has not. Remember, a combination of auditory and visual aids is the best way to convey your ideas.

General guidelines

- Do not merely display data. It is important to show the implications and significance of the work.
- All display materials should be easily readable at a distance of about 5 feet. If this appears impossible for a specific item, then handout should be used to convey that particular information.
- Determine a logical and attractive layout for your poster in advance. If the poster is composed in sections, number the sections so that they can be easily remounted at the conference site.
- Find out your poster session number and booth location in advance and verify before mounting your poster at the site.
- Display the title, author, and affiliation in large-enough letters at the top of the poster board.
- Take with you useful items such as sketch pads and marking pens.
- Make sure your materials are sufficiently lightweight and thin to be mounted on a vertical board. For heavy and thick materials, consider using Velcro material (available in fabric stores) for mounting the items.
- Do not write on display boards.
- Attempt to have your poster completely set up at least 15 minutes before your presentation. Take your time to remove all your materials from the poster board at the conclusion of the display.
- During the presentation, discuss your topic conversationally rather than lecturing or simply reading a summary. The discussion may begin with a question from an interested attendee. You may initiate a discussion by pointing out the figure that depicts the conclusions of your research. Questions and explanations can then follow from that point.

Poster evaluation criteria

A poster session can be rated on an ordinal scale from excellent to inadequate (excellent, good, fair, poor, inadequate). Some of the specific criteria are:

1. Poster is well designed, organized, and attractive.
2. Major points of the work are presented.
3. Diagrams present enough materials to clearly identify: Problem, Objective, Methodology, Data Analysis, Results, and Conclusions.
4. Oral presentation/explanations are clear and understandable.
5. Handout materials are available and useful.

Appendix 13: Project-relevant quotes

A business that makes nothing but money is a poor business.

Henry Ford

A plan that will not fit on one page cannot be understood.

Mark Ardis

A good plan can help with risk analyses but it will never guarantee the smooth running of the project.

Bentley and Borman

A plan is the map of the wise.

Adedeji B. Badiru

A project is complete when it starts working for you, rather than you working for it.

Scott Allen

A project without a critical path is like a ship without a rudder.

D. Meyer

A task is not done until it is done.

Louis Frie

A well-constructed project management workshop should give people a solid foundation to build on.

Bentley and Borman

Add little to little and there will be a big pile.

Ovid

All things are created twice; first mentally; then physically. The key to creativity is to begin with the end in mind, with a vision and a blue print of the desired result.

Stephen Covey

An error does not become truth by reason of multiplied propagation, nor does truth become error because nobody sees it.

Mahatma Gandhi

An intelligent person armed with a checklist is no substitute for experience.

Joy Gumz

An ounce of action is worth a ton of theory.

Friedrich Engels

Any idiot can point out a problem .... A leader is willing to do something about it! Leaders solve problems!

Tony Robbins

Any fool can criticize, condemn, and complain, and most fools do. But it takes character and self-control to be understanding and forgiving.

Dale Carnegie

As has been taught to teachers of the Harvard Business School, the art of asking good questions is often the most important element of managerial tasks.

Parte Bose

Assumption is the mother of all screw-ups.

Wethern's Law of Suspended Judgement

At times, project managers seem to forget that many of the conventional forms, charts, and tables that they must fill out are intended to serve as aids, not punishments.

Mantel, Meredith, Shafer, and Sutton

Before anything can be repeatable or reusable, it must be usable.

Woody Williams

Being a Project Manager is like being an artist, you have the different colored process streams combining into a work of art

Greg Cimmarrusti

Believe and act as if it were impossible to fail.

Unknown

Beware the time-driven project with an artificial deadline.

M. Dobson

Change is not made without inconvenience, even from worse to better.

Samuel Johnson

Communication is the root of everything else.

Adedeji B. Badiru

Data is like garbage. You'd better know what you are going to do with it before you collect it.

Mark Twain

Divide and conquer is the way to get projects done.

Adedeji B. Badiru

Do not repeat the tactics which have gained you one victory, but let your methods be regulated by the infinite variety of circumstances.

Sun Tzu

Do not squander time, for that is the stuff life is made of.

Benjamin Franklin

Don't do anything you don't have to do.

Louis Fried

Don't use a sledgehammer to crack a walnut, but equally don't agree important things informally where there is a chance of a disagreement later over what was agreed.

Colin Bentley

Each completed task establishes certain parameters and imposes constraints on the next task.

Louis Fried

Effective leaders help others to understand the necessity of change and to accept a common vision of the desired outcome.

John Kotter

Ensure your documentation is short and sharp and make much more use of people-to-people communication.

Bentley and Borman

Even if you are on the right track, you will get run over if you just sit there.

Will Rogers

Event management is the same as for any project— the project plan needs to include an appropriate change control process.

Brenda Treasure

Every moment is a golden one for him who has the vision to recognize it as such.

Henry Miller

Every person takes the limits of their own field of vision for the limits of the world.

Arthur Schopenhauer

External audits are routine in the financial area. I fail to understand why nonprofits don't use them more in the vital program area.

E. Stoesz

First comes thought; then organization of that thought, into ideas and plans; then transformation of those plans into reality. The beginning, as you will observe, is in your imagination.

Napoleon Hill

For a project plan to be effective it must equally address the parameters of "activity time" and "activity logic." This logical relationship is required to model the effect schedule variance will have downstream in the project.

Rory Burke

Get it done and put it behind you.

Adedeji B. Badiru

Get the right people. Then no matter what all else you might do wrong after that, the people will save you. That's what management is all about.

Tom DeMarco

Getting more things done requires focusing on fewer things to do.

Adedeji B. Badiru

Give me six hours to chop down a tree and I will spend the first four sharpening the axe.

Abraham Lincoln

Good business leaders create a vision, articulate the vision, passionately own the vision, and relentlessly drive it to completion.

Jack Welch

Good judgment comes from experience, and experience comes from bad judgment.

Fred Brooks

Good leaders do not take on all the work themselves; neither do they take all the credit.

Woody Williams

Grass is always greener where you most need it to be dead.

Adedeji B. Badiru

Great Works are performed not by Strength, but perseverance.

Samuel Johnson (English author, 1709–1784)

Haste makes waste just as rush makes crash.

Adedeji B. Badiru

He has half the deed done who has made a beginning.

Horace

He who has a "why" to live for can bear with almost any "how."

Friedrich Nietzsche

I don't know the key to success, but the key to failure is trying to please everyone

Bill Cosby

I have witnessed boards that continued to waste money on doomed projects because no one was prepared to admit they were failures, take the blame and switch course. Smaller outfits are more willing to admit mistakes and dump bad ideas.

Luke Johnson

I hate housework! You do the dishes, and six months later you have to start all over again.

Joan Rivers

I like work. It fascinates me. I can sit and look at it for hours.

Jerome Klapka Jerome

I think there is something, more important than believing: Action! The world is full of dreamers, there aren't enough who will move ahead and begin to take concrete steps to actualize their vision.

W. Clement Stone

I never dreamed about success. I worked for it.

Estee Lauder

If an IT project works the first time, it was a very small and simple project.

Cornelius Fitchner

If an IT project works the first time, it was in your nightly dreams. Time to wake up and get to work.

Cornelius Fitchner

If everyone is thinking alike, someone isn't thinking.

General George Patton Jr.

If everything seems under control, you're not going fast enough.

Mario Andretti

If I have seen farther than others, it is because I was standing on the shoulders of giants.

Isaac Newton

If it is not documented, it doesn't exist. As long information is retained in someone's head, it is vulnerable to loss.

Louis Fried

If money could just grow on trees, every miscreant would have some.

Adedeji B. Badiru

If you always blame others for your mistakes, you will never improve.

Joy Gumz

If you can't describe what you are doing as a process, you don't know what you're doing.

W. Edwards Deming

If you don't start, it's certain you won't arrive.

Zig Ziglar

If you have never recommended canceling a project, you haven't been an effective project manager.

Woody Williams

If you wait long enough, you won't have to buy new technology.

Adedeji Badiru

If, on your team, everyone's input is not encouraged, valued, and welcome, why call it a team?

Woody Williams

Imagination is more important than knowledge.

Albert Einstein

In NASA, we never punish error. We only punish the concealment of error.

Al Siepert

In poorly run projects, problems can go undetected until the project fails. It's like the drip ... drip ... drip of an leaky underground pipe. Money is being lost, but you don't see it until there is an explosion.

Joy Gumz

Innovative efforts should never report to line managers charged with responsibility for ongoing operations. The new project is an infant and will remain one for the foreseeable future, and infants belong in the nursery. The "adults," that is, the executives in charge of existing businesses or products will have neither the time nor understanding for the infant.

Peter Drucker

It doesn't take a lot of salt to add the flavor.

Unknown

It does not matter how slowly you go, so long as you do not stop.

Confucius

It is always easier to talk about change than to make it.

Alvin Toffler

It is better to know some of the questions than all of the answers.

James Thurber

It is not a question of how well each process works; the question is how well they all work together.

Lloyd Dobyns and Clare Crawford-Mason

It is useless to desire more time if you are already wasting what little you have.

James Allen

It must be considered that there is nothing more difficult to carry out nor more doubtful of success nor more dangerous to handle than to initiate a new order of things.

Machiavelli

It's easy to see, hard to foresee.

Benjamin Franklin

It's not enough that we do our best; sometimes we have to do what's required.

Winston Churchill

Keep your dreams alive. Understand to achieve anything requires faith and belief in yourself, vision, hard work, determination, and dedication. Remember all things are possible for those who believe.

Gail Devers

Know when to cut your losses if necessary. Don't let your desire to succeed be the enemy of good judgment. If Napoleon had left Moscow immediately, he may have returned with a salvageable army.

Jerry Manas

Lavish credit on anyone and everyone who helped you the least bit.

Tom Peters

Let a plan be your project's guiding light.

Adedeji B. Badiru

Life is like riding a bicycle. In order to keep your balance, you must keep moving.

Albert Einstein

Like organic entities, projects have life cycles. From a slow beginning they progress to a buildup of size, then peak, begin a decline, and finally must be terminated. (Also, like other organic entities, they often resist termination.)

Meredith and Mantel

Make an extensive table of project "deliverables." Label one column "as requested." Create another column labeled "could be." Make each "could be" wild and woolly!

Tom Peters

Management is doing things right; leadership is doing the right things.

Peter F. Drucker

Momentum is a fragile force. Its worst enemy: procrastination. Its best friend: a deadline (think Election Day). Implication no. 1 (and there is no no. 2): Get to work! NOW!

Tom Peters

Most success depends on colleagues, on the team. People at the top have large egos, but you must never say "I": it's always "we."

Frank Lampl

My personal philosophy is not to undertake a project unless it is manifestly important and nearly impossible.

Edwin Land

Never allow a person to tell you no who doesn't have the power to say yes.

Eleanor Roosevelt

No major project is ever installed on time, within budget, with the same staff that started it.

Edwards, Butler, Hill, and Russell

No matter how good the team or how efficient the methodology, if we're not solving the right problem, the project fails.

Woody Williams

No one can whistle a symphony. It takes a whole orchestra.

H.E. Luccock

Nobody knows how Honda is organized, except that it uses lots of project teams and is quite flexible.

Kenichi Ommae

Of all the things I've done, the most vital is coordinating the talents of those who work for us and pointing them towards a certain goal.

Walt Disney

Operations keeps the lights on, strategy provides a light at the end of the tunnel, but project management is the train engine that moves the organization forward.

Joy Gumz

Our focus on meaningful projects means serving a useful social purpose while generating high-quality profits.

Masimi Iijima

People are more inclined to be drawn in if their leader has a compelling vision. Great leaders help people get in touch with their own aspirations and then will help them forge those aspirations into a personal vision.

John Kotter

People buy into the leader before they buy into the vision.

John C. Maxwell

Pharmaceutical projects are like fresh fruit—they depreciate if they are not tended to, and they do poorly if sitting on the shelf with long periods of inactivity.

R. Burns

Planning without action is futile, action without planning is fatal.

Cornelius Fitchner

Plans are nothing; planning is everything.

Dwight D. Eisenhower

Plans are only good intentions unless they immediately degenerate into hard work.

Peter Drucker

Plans are worthless, but planning is invaluable.

Peter Drucker

PMs are the most creative pros in the world; we have to figure out everything that could go wrong, before it does.

Fredrik Haren

Price is what you pay; value is what you get.

Warren Buffet

Process for process sake is not good for goodness sake.

Lynn A. Edmark

Project management can be defined as a way of developing structure in a complex project, where the independent variables of time, cost, resources and human behavior come together.

Rory Burke

Project management is like juggling three balls—time, cost and quality. Program management is like a troupe of circus performers standing in a circle, each juggling-three balls and swapping balls from time to time.

G. Reiss

Project management is the art of creating the illusion that any outcome is the result of a series of predetermined, deliberate acts when, in fact, it was dumb luck.

Harold Kerzner

Project managers function as bandleaders who pull together their players each a specialist with individual score and internal rhythm. Under the leader's direction, they all respond to the same beat.

L.R. Sayles

Project managers rarely lack organizational visibility, enjoy considerable variety in their day- to-day duties, and often have the prestige associated with work on the enterprise's high- priority objectives.

Meredith and Mantel

Project proposals, business cases or cost benefit analyses are probably being massaged (either by underestimating costs or timeframes or by being very optimistic about the benefits) so projects will be approved.

Bentley and Borman

Projects progress quickly until they become 90% complete; then remain at 90% complete forever.

Edwards, Butler, Hill, and Russell

Reconnaissance memoranda should always be written in the simplest style and be purely descriptive. They should never stray from their objective by introducing extraneous ideas.

Napoleon Bonaparte

Resource is the engine of performance.

Adedeji B. Badiru

Rewards and motivation are an oil change for project engines. Do it regularly and often.

Woody Williams

Running a project without a WBS is like going to a strange land without a roadmap.

J. Phillips

Skills can be learned while experience must be earned.

Joy Gumz

Some of the most flowery praise you hear on the subject of teams is only hypocrisy. Managers learn to talk a good game about teams even when they're secretly threatened by the whole concept.

Tom DeMarco

Some things are better done than described.

Hunt and Thomas

Tell me and I forget; show me and I remember; involve me and I understand.

Chinese Proverb

Testing proves the presence of bugs but not their absence.

Woody Williams

The conditions attached to a promise are forgotten and the promise is remembered.

Edwards, Butler, Hill, and Russell

The functional groups should not be allowed to stretch out the project for the sake of improvement, refinement, or the investigation of the most remote potential risk.

Meredith and Mantel

The P in PM is as much about "people manage-ment" as it is about "project management."

Cornelius Fichtner

The project manager is expected to integrate all aspects of the project, ensure that the proper knowl-edge and resources are available when and where needed, and above all, ensure that the expected results are produced in a timely, cost- effective manner.

Meredith and Mantel

The project manager must be able to develop a fully integrated information and control system to plan, instruct, monitor and control large amounts of data, quickly and accurately to facilitate the problem-solving and decision-making process.

Rory Burke

The real problem is what to do with problem solvers after the problem is solved.

Gay Talese

The reasonable man adapts himself to the world; the unreasonable one persists in trying to adapt the world to himself. Therefore, all progress depends on the unreasonable man.

George Bernard Shaw

The testers won't break the system but the user who thinks the cd-rom drive as a drinks holder will.

Cornelius Fitchner

The urgent problems are seldom the important ones.

Dwight D. Eisenhower

There are two types of software: bad software and the next release.

Cornelius Fitchner

There is nothing more perishable than an airline seat—unless it is time on a project.

Joy Gumz

There is time for everything, and a season for each activity.

Ecclesiastes 3:1

Things which matter most must never be at the mercy of things which matter least.

Goethe

Think twice and act wise.

Unknown

This ad hoc approach to project management—coupled as it frequently is, with an on-the-job training philosophy—is pervasive. It is also pernicious.

Jack Meredith

Time is like water in a sponge, the more you squeeze, the more you may get.

Chinese saying

To get a project off the ground, tell a colleague it was their idea. They will put their heart and soul into making it successful.

T. Wouhra

To get more done, try and do less.

Adedeji B. Badiru

True motivation comes from achievement, personal development, job satisfaction, and recognition.

Frederick Herzberg

Trying to manage a project without project management is like trying to play a football game without a game plan.

K. Tate

Unengaged sponsor sinks the ship.

Angela Waner

Unless commitment is made, there are only promises and hopes … but no plans.

Peter Drucker

Vision is the art of seeing what is invisible to others.

Jonathan Swift

Vision without action is a dream. Action without vision is simply passing the time. Action with vision is making a positive difference.

Joel Barker

We are what we repeatedly do. Excellence, therefore, is not an act but a habit.

Aristotle

We cannot drive people; we must direct their development. Teach and lead.

Woody Williams

We trained hard, but it seemed that every time we were beginning to form into teams we would be reorganized. I was to learn later in life that we tend to meet any new situation by reorganizing; and what a wonderful method it can be for creating the illusion of progress while producing confusion, inefficiency and demoralization.

Petronius Arbiter, 210 BC

We will either find a way, or make one.

Hannibal

What is not on paper has not been said.

Anonymous

What is truth today may be falsehood tomorrow. Never confuse your plan with truth.

Woody Williams

What we are looking for is managers who are awake enough to alter the world as they find it, to make it harmonize with what they and their people are trying to accomplish.

Tom DeMarco

What we learn from lessons learned is that we don't learn from lessons learned.

T. Block

Whatever we do must be in accord with human nature. We cannot drive people; we must direct their development. The general policy of the past has been to drive; but the era of force must give way to the era of knowledge, and the policy of the future will be to teach and lead, to the advantage of all concerned.

Henry L. Gantt

When a risk occurs, with some ingenuity, this may open up an opportunity, and conversely when pursuing an opportunity there will be associated risks. Risks are generally deemed acceptable if the possible gains exceed the possible losses.

Rory Burke

When debugging, novices insert corrective code; experts remove defective code.

Richard Pattis

When end users get involved in the final stages of testing, light bulbs go on, and they often have an "aha" moment. Unfortunately, that is often too late.

Frank R. Parth

When I dare to be powerful—to use my strength in the service of my vision, then it becomes less and less important whether I am afraid.

Audre Lorde

When the territory and the map disagree, believe the territory.

Swiss Army Manual

Whilst you can practice good project management without EVM, you cannot practice EVM effectively without good project management.

Steve Crowther

Why do so many professionals say they are project managing, when what they are actually doing is fire fighting?

Colin Bentley

With vision there is no room to be frightened. No reason for intimidation. It's time to march forward! Let's be confident and positive!

Charles R. Swindoll

Working ten hour days allows you to fall behind twice as fast as you could working five hour days.

Issac Assimov

You can only elevate individual performance by elevating that of the entire system.

W. Edwards Deming

You can't keep it all in your head. Project control tools are an absolute necessity for the control of large projects.

Louis Fried

You can't turn a herd of turtles into a twenty-mule work team.

L. Todryk

You may con a person into committing to an unreasonable deadline, but you cannot bully them into meeting it.

Edwards, Butler, Hill, and Russell

You must learn from the mistakes of others. You can't possibly live long enough to make them all yourself.

Sam Levenson

Appendix 14: Conversion factors and expressions

Numbers and prefixes

yotta (10^{24})	1 000 000 000 000 000 000 000 000
zetta (10^{21})	1 000 000 000 000 000 000 000
exa (10^{18})	1 000 000 000 000 000 000
peta (10^{15})	1 000 000 000 000 000
tera (10^{12})	1 000 000 000 000
giga (10^{9})	1 000 000 000
mega (10^{6})	1 000 000
kilo (10^{3})	1 000
hecto (10^{2})	100
deca (10^{1})	10
deci (10^{-1})	0.1
centi (10^{-2})	0.01
milli (10^{-3})	0.001
micro (10^{-6})	0.000 001
nano (10^{-9})	0.000 000 001
pico (10^{-12})	0.000 000 000 001
femto (10^{-15})	0.000 000 000 000 001
atto (10^{-18})	0.000 000 000 000 000 001
zepto (10^{-21})	0.000 000 000 000 000 000 001
yocto (10^{-24})	0.000 000 000 000 000 000 000 001
stringo (10^{-35})	0.000 000 000 000 000 000 000 000 000 000 000 01

Constants

Speed of light	2.997925×10^{10} cm/s
	983.6×10^6 ft/s
	186,284 mi/s
Velocity of sound	340.3 m/s
	1116 ft/s
Gravity	9.80665 m/s^2
(acceleration)	32.174 ft/s^2
	386.089 in/s^2

Area

Multiply	By	To obtain
acres	43,560	square feet
	4,047	square meters
	4,840	square yards
	0.405	hectares
square centimeters	0.155	square inches
square feet	144	square inches
	0.09290	square meters
	0.1111	square yards
square inches	645.16	square millimeters
square kilometers	0.3861	square miles
square meters	10.764	square feet
	1.196	square yards
square miles	640	acres
	2.590	square kilometers

Volume

Multiply	By	To obtain
acre-feet	1233.5	cubic meters
cubic centimeters	0.06102	cubic inches
cubic feet	1728	cubic inches
	7.480	gallons (US)
	0.02832	cubic meters
	0.03704	cubic yards
liters	1.057	liquid quarts
	0.908	dry quarts
	61.024	cubic inches
gallons (US)	231	cubic inches
	3.7854	liters
	4	quarts
	0.833	British gallons
	128	U.S. fluid ounces
quarts (US)	0.9463	liters

Energy, heat power

Multiply	By	To obtain
BTU	1055.9	joules
	0.2520	kilogram-calories
watt-hour	3600	joules
	3.409	BTU
HP (electric)	746	watts
BTU/second	1055.9	watts
watt-second	1.00	joules

Mass

Multiply	By	To obtain
carats	0.200	cubic grams
grams	0.03527	ounces
kilograms	2.2046	pounds
ounces	28.350	grams
pounds	16	ounces
	453.6	grams
stones (UK)	6.35	kilograms
	14	pounds
tons (net)	907.2	kilograms
	2000	pounds
	0.893	gross tons
	0.907	metric tons
tons (gross)	2240	pounds
	1.12	net tons
	1.016	metric tons
tonnes (metric)	2204.623	pounds
	0.984	gross pounds
	1000	kilograms

Temperature

Conversion formulas	
Celsius to Kelvin	$K = C + 273.15$
Celsius to Fahrenheit	$F = (9/5)C + 32$
Fahrenheit to Celsius	$C = (5/9)(F - 32)$
Fahrenheit to Kelvin	$K = (5/9)(F + 459.67)$
Fahrenheit to Rankin	$R = F + 459.67$
Rankin to Kelvin	$K = (5/9)R$

Velocity

Multiply	By	To obtain
feet/minute	5.080	millimeters/second
feet/second	0.3048	meters/second
inches/second	0.0254	meters/second
kilometers/hour	0.6214	miles/hour
meters/second	3.2808	feet/second
	2.237	miles/hour
miles/hour	88.0	feet/minute
	0.44704	meters/second
	1.6093	kilometers/hour
	0.8684	knots
knot	1.151	miles/hour

Pressure

Multiply	By	To obtain
atmospheres	1.01325	bars
	33.90	feet of water
	29.92	inches of mercury
	760.0	millimeters of mercury
bars	75.01	centimeters of mercury
	14.50	pounds/square inch
dynes/square centimeter	0.1	newtons/square meter
newtons/square centimeter	1.450	pounds/square inch
pounds/square inch	0.06805	atmospheres
	2.036	inches of mercury
	27.708	inches of water
	68.948	millibars
	51.72	millimeters of mercury

Distance

Multiply	By	To obtain
angstroms	10^{-10}	meters
feet	0.30480	meters
	12	inches
inches	25.40	millimeters
	0.02540	meters
	0.08333	feet
kilometers	3280.8	feet
	0.6214	miles
	1094	yards
meters	39.370	inches
	3.2808	feet
	1.094	yards
miles	5280	feet
	1.6093	kilometers
	0.8694	nautical miles
millimeters	0.03937	inches
nautical miles	6076	feet
	1.852	kilometers
yards	0.9144	meters
	3	feet
	36	inches

Physical relationships

$$D = \frac{m}{V}$$

D: density
m: mass
V: volume
$$\left(\frac{g}{cm^3} = \frac{kg}{m^3}\right)$$

$$d = v \cdot t$$

d: distance (m)
v: velocity (m/s)
t: time (s)

$$a = \frac{vf - vi}{t}$$

a: acceleration (m/s²)
vf: final velocity (m/s)
vi: initial velocity (m/s)
t: time (s)

$$d = vi \cdot t + \frac{1}{2} \cdot a \cdot t^2$$

d: distance (m)
vi: initial velocity (m/s)
t: time (s)
a: acceleration (m/s²)

$$F = m \cdot a$$

F: net force (N)
m: mass (kg)
a: acceleration (m/s²)

$$P = \frac{W}{t}$$

P: power (W)
W: work (J)
t: time (s)

$$K.E. = \frac{1}{2} \cdot m \cdot v^2$$

K.E.: kinetic energy
m: mass (kg)
v: velocity (m/s)

$$Fe = \frac{kQ_1Q_2}{d^2}$$

Fe: electrical force (N)
k: Coulomb's constant
$$\left(k = 9 \times 10^9 \frac{N \cdot m^2}{c^2}\right)$$
$Q_1 \cdot Q_2$: electrical charges (C)
d: separation distance (m)

$$V = \frac{W}{Q}$$

V: electrical potential difference (V)
W: work done (J)
Q: electric charge moving (C)

(Continued)

$$Fg = \frac{G \cdot m_1 \cdot m_2}{d^2}$$

Fg: force of gravity (N)
G: universal gravitational constant
$\left(G = 6.67 \times 10^{-11} \dfrac{N\,m^2}{kg^2}\right)$

m_1, m_2: masses of the two objects (kg)
d: separation distance (m)

$$p = m \cdot v$$

p: momentum (kg · m/s)
m: mass
v: velocity

$$W = F \cdot d$$

W: work (J)
F: force (N)
d: distance (m)

$$I = \frac{Q}{t}$$

I: electric current (A)
Q: electric charge flowing (C)
t: time (s)

$$W = V.I.t$$

W: electrical energy (J)
V: voltage (V)
I: current (A)
t: time (s)

$$P = V \cdot I$$

P: power (W)
V: voltage (V)
I: current (A)

$$H = c \cdot m \cdot \Delta T$$

H: heat energy (J)
m: mass (kg)
ΔT: change in temperature (°C)
c: specific heat (J/kg · °C)

Units of measurement

English system		Metric system		
1 foot (ft)	=12 inches (in) 1′ = 12″	mm	millimeter	0.001 m
1 yard (yd)	=3 feet	cm	centimeter	0.01 m
1 mile (mi)	=1760 yards	dm	decimeter	0.1 m
1 sq. foot	=144 sq. inches	m	meter	1 m
1 sq. yard	=9 sq. feet	dam	decameter	10 m
1 acre	=4840 sq. yards = 43,560 ft²	hm	hectometer	100 m
1 sq. mile	=640 acres	km	kilometer	1000 m

Note: Prefixes also apply to l (liter) and g (gram).

Common notations

Units of measurement	Abbreviation	Relation
meter	m	Length
hectare	ha	Area
tonne	t	Mass
kilogram	kg	Mass
nautical mile	M	Distance (navigation)
knot	kn	Speed (navigation)
liter	L	Volume or capacity
second	s	Time
hertz	Hz	Frequency
candela	cd	Luminous intensity
degree celsius	°C	Temperature
kelvin	K	Thermodynamic temperature
pascal	Pa	Pressure, stress
joule	J	Energy, work
newton	N	Force
watt	W	Power, radiant flux
ampere	A	Electric current
volt	V	Electric potential
ohm	Ω	Electric resistance
coulomb	C	Electric charge

Kitchen conversion units

A pinch	1/8 tsp or less
3 tsp	1 tbsp
2 tbsp	1/8 c
4 tbsp	1/4 c
16 tbsp	1 c
5 tbsp + 1 tsp	1/3 c
4 oz	1/2 c
8 oz	1 c
16 oz	1 lbs
1 oz	2 tbsp fat or liquid
1 c of liquid	1/2 pt
2 c	1 pt
2 pt	1 qt
4 c of liquid	1 qt
4 qts	1 gallon
8 qts	1 peck (such as apples, pears, etc.)
1 jigger	1½ fl oz
1 jigger	3 tbsp

Appendix 15: Glossary of project management terms

- *ABC* (activity-based costing). Bottom-up estimating and summation based on material and labor required for activities making up a project.
- *Activity.* A component of work performed during the course of a project.
- *Activity duration.* The time in calendar units between the start and finish of a schedule activity.
- *Activity resource estimating.* The process of estimating the types and quantities of resources required to perform each schedule activity.
- *Activity sequencing.* The process of identifying and documenting dependencies among schedule activities.
- *Authority.* The right to apply project resources, expend funds, make decisions, or give approvals.
- *Bar chart.* A graphic display of schedule-related information. In a typical bar chart, schedule activities or work breakdown structure components are listed on the left side of the chart, dates are shown across the top, and activity durations are shown as date-placed horizontal bars. Also called a Gantt chart.
- *Baseline.* The approved time-phased plan (for a project, a work breakdown structure component, a work package, or a schedule activity), plus or minus approved project scope, cost, schedule, and technical changes. Generally refers to the current baseline, but may refer to the original or some other baseline. Usually used with a modifier (e.g., cost baseline, schedule baseline, performance measurement baseline, technical baseline).
- *Best practices.* Processes, procedures, and techniques that have consistently demonstrated achievement of expectations and that are documented for the purposes of sharing, repetition, replication, adaptation, and refinement.

- *Baseline start date.* The start date of a schedule activity in the approved schedule baseline.
- *Change control.* Identifying, documenting, approving or rejecting, and controlling changes to the project baselines.
- *Close project.* The process of finalizing all activities across all of the project process groups to formally close the project or phase.
- *Common cause.* A source of variation that is inherent in the system and predictable. On a control chart, it appears as part of the random process variation (i.e., variation from a process that would be considered normal or not unusual), and is indicated by a random pattern of points within the control limits. Also referred to as random cause. Contrast with special cause.
- *Configuration management system.* A subsystem of the overall project management system. It is a collection of formal documented procedures used to apply technical and administrative direction and surveillance to: identify and document the functional and physical characteristics of a product, result, service, or component; control any changes to such characteristics; record and report each change and its implementation status; and support the audit of the products, results, or components to verify conformance to requirements. It includes the documentation, tracking systems, and defined approval levels necessary for authorizing and controlling changes. In most application areas, the configuration management system includes the change control system.
- *Constraint.* The state, quality, or sense of being restricted to a given course of action or inaction. An applicable restriction or limitation, either internal or external to the project, that will affect the performance of the project or a process. For example, a schedule constraint is any limitation or restraint placed on the project schedule that affects when a schedule activity can be scheduled and is usually in the form of fixed imposed dates. A cost constraint is any limitation or restraint placed on the project budget such as funds available over time. A project resource constraint is any limitation or restraint placed on resource usage, such as what resource skills or disciplines are available and the amount of a given resource available during a specified time frame.
- *Contingency reserve.* The amount of funds, budget, or time needed above the estimate to reduce the risk of overruns of project objectives to a level acceptable to the organization.
- *Control.* Comparing actual performance with planned performance, analyzing variances, assessing trends to effect process improvements, evaluating possible alternatives, and recommending appropriate corrective action as needed.

- *Control chart.* A graphic display of process data over time and against established control limits, and that has a centerline that assists in detecting a trend of plotted values toward either control limit.
- *Control limits.* The area composed of three standard deviations on either side of the centerline, or mean, of a normal distribution of data plotted on a control chart that reflects the expected variation in the data.
- *Cost control.* The process of influencing the factors that create variances, and controlling changes to the project budget.
- *Cost of quality (COQ).* Determining the costs incurred to ensure quality. Prevention and appraisal costs (cost of conformance) include costs for quality planning, quality control (QC), and quality assurance to ensure compliance to requirements (i.e., training, QC systems, etc.). Failure costs (cost of nonconformance) include costs to rework products, components, or processes that are noncompliant, costs of warranty work and waste, and loss of reputation.
- *Cost performance index (CPI).* A measure of cost efficiency on project. It is the ratio of earned value (EV) to actual costs (AC). CPI = EV divided by AC. A CPI value equal to or greater than one indicates a favorable condition and a value less than one indicates an unfavorable condition.
- *Cost-plus-fee (CPF).* A type of cost-reimbursable contract where the buyer reimburses the seller for seller's allowable costs for performing the contract work and seller also receives a fee calculated as an agreed-upon percentage of the costs. The fee varies with the actual cost.
- *Cost-plus-fixed-fee (CPFF) contract.* A type of cost-reimbursable contract where the buyer reimburses the seller for the seller's allowable costs (allowable costs are defined by the contract) plus a fixed amount of profit (fee).
- *Cost-plus-incentive-fee (CPIF) contract.* A type of cost-reimbursable contract where the buyer reimburses the seller for the seller's allowable costs (allowable costs are defined by the contract), and the seller earns its profit if it meets defined performance criteria.
- *Cost-plus-percentage of cost (CPPC).* See *cost-plus-fee.*
- *Cost-reimbursable contract.* A type of contract involving payment (reimbursement) by the buyer to the seller for the seller's actual costs, plus a fee typically representing seller's profit. Costs are usually classified as direct costs or indirect costs. Direct costs are costs incurred for the exclusive benefit of the project, such as salaries of full-time project staff. Indirect costs, also called overhead and general and administrative cost, are costs allocated to the project by the performing organization as a cost of doing business, such as salaries of management indirectly involved in the project, and

cost of electric utilities for the office. Indirect costs are usually calculated as a percentage of direct costs. Cost-reimbursable contracts often include incentive clauses where, if the seller meets or exceeds selected project objectives, such as schedule targets or total cost, then the seller receives from the buyer an incentive or bonus payment.

- *Cost variance (CV)*. A measure of cost performance on a project. It is the algebraic difference between earned value (EV) and actual cost (AC). CV = EV minus AC. A positive value indicates a favorable condition and a negative value indicates an unfavorable condition.
- *Crashing*. A specific type of project schedule compression technique performed by taking action to decrease the total project schedule duration after analyzing a number of alternatives to determine how to get the maximum schedule duration compression for the least additional cost. Typical approaches for crashing a schedule include reducing schedule activity durations and increasing the assignment of resources on schedule activities. See also fast tracking.
- *Create WBS (work breakdown structure)*. The process of subdividing the major project deliverables and project work into smaller, more manageable components.
- *Critical activity*. Any schedule activity on a critical path in a project schedule. Most commonly determined by using the critical path method. Although some activities are "critical," in the dictionary sense, without being on the critical path, this meaning is seldom used in the project context.
- *Critical chain method*. A schedule network, analysis technique that modifies the project schedule to account for limited resources. The critical chain method mixes deterministic and probabilistic approaches to schedule network analysis.
- *Critical path*. Generally, but not always, the sequence of schedule activities that determines the duration of the project. Generally, it is the longest path through the project. However, a critical path can end, as an example, on a schedule milestone that is in the middle of the project schedule and that has a finish-no-later-than imposed date schedule constraint. See also critical path method.
- *Critical path method (CPM)*. A schedule network analysis technique used to determine the amount of scheduling flexibility (the amount of float) on various logical network paths in the project schedule network, and to determine the minimum total project duration. Early start and finish dates are calculated by means of a forward pass, using a specified start date. Late start and finish dates are calculated by means of a backward pass, starting from a specified completion date, which sometimes is the project early finish date determined during the forward pass calculation.

- *Decision tree analysis.* The decision tree is a diagram that describes a decision under consideration and the implications of choosing one or another of the available alternatives. It is used when some future scenarios or outcomes of actions are uncertain. It incorporates probabilities and the costs or rewards of each logical path of events and future decisions, and uses expected monetary value analysis to help the organization identify the relative values of alternate actions. See also expected monetary value analysis.
- *Decomposition.* A planning technique that subdivides the project scope and project deliverables into smaller, more manageable components, until the project work associated with accomplishing the project scope and providing the deliverables is defined in sufficient detail to support executing, monitoring, and controlling the work.
- *Defect.* An imperfection or deficiency in a project component where that component does not meet its requirements or specifications and needs to be either repaired or replaced.
- *Defect repair.* Formally documented identification of a defect in a project component with a recommendation to either repair the defect or completely replace the component.
- *Deliverable.* Any unique and verifiable product, result, or capability to perform a service that must be produced to complete a process, phase, or project. Often used more narrowly in reference to an external deliverable, which is a deliverable that is subject to approval by the project sponsor or customer.
- *Delphi technique.* An information-gathering technique used as a way to reach a consensus of experts on a subject. Experts on the subject participate in this technique anonymously. A facilitator uses a questionnaire to solicit ideas about important project points related to the subject. The responses are summarized and are then recirculated to the experts for further comment. Consensus may be reached in a few rounds of this process. The Delphi technique helps reduce bias in the data and keeps any one person from having undue influence on the outcome.
- *Develop project charter.* The process of developing the project charter that formally authorizes a project.
- *Discrete effort.* Work effort that is directly identifiable to the completion of specific work breakdown structure components and deliverables, and that can be directly planned and measured. Contrast with apportioned effort.
- *Dummy activity.* A schedule activity of zero duration used to show a logical relationship in the arrow diagramming method. Dummy activities are used when logical relationships cannot be completely or correctly described with schedule activity arrows. Dummy activities are generally shown graphically as a dashed line headed by an arrow.

- *Early finish date (EF).* In the critical path method, the earliest possible point in time at which the uncompleted portions of a schedule activity (or the project) can finish, based on the schedule network, logic, the data date, and any schedule constraints. Early finish dates can change as the project progresses and as changes are made to the project management plan.
- *Early start date (ES).* In the critical path method, the earliest possible point in time at which the uncompleted portions of a schedule activity (or the project) can start, based on the schedule network logic, the data date, and any schedule constraints. Early start dates can change as the project progresses and as changes are made to the project management plan.
- *Earned value (EV).* The value of completed work expressed in terms of the approved budget assigned to that work for a schedule activity or work breakdown structure component. Also referred to as the budgeted cost of work performed (BCWP).
- *Earned value management (EVM).* A management methodology for integrating scope, schedule, and resources, and for objectively measuring project performance and progress. Performance is measured by determining the budgeted cost of work performed (i.e., earned value) and comparing it to the actual cost of work performed (i.e., actual cost). Progress is measured by comparing the earned value to the planned value.
- *Earned value technique (EVT).* A specific technique for measuring the performance of work for a work breakdown structure component, control account, or project. Also referred to as the earning rules and crediting method.
- *Effort.* The number of labor units required to complete a schedule activity or work breakdown structure component. Usually expressed as staff hours, staff days, or staff weeks. Contrast with duration.
- *Enterprise.* A company, business, firm, partnership, corporation, or governmental agency.
- *Enterprise environmental factors.* Any or all external environmental factors and internal organizational environmental factors that surround or influence the project's success. These factors are from any or all of the enterprises involved in the project, and include organizational culture and structure, infrastructure, existing resources, commercial databases, market conditions, and project management software.
- *Execute.* Directing, managing, performing, and accomplishing the project work, providing the deliverables, and providing work performance information.
- *Expected monetary value (EMV) analysis.* A statistical technique that calculates the average outcome when the future includes scenarios

that may or may not happen. A common use of this technique is within decision tree analysis. Modeling and simulation are recommended for cost and schedule risk analysis because it is more powerful and less subject to misapplication than expected monetary value analysis.

- *Expert judgment.* Judgment provided based upon expertise in an application area, knowledge area, discipline, industry, etc. as appropriate for the activity being performed. Such expertise may be provided by any group or person with specialized education, knowledge, skill, experience, or training, and is available from many sources, including other units within the performing organization; consultants; stakeholders, including customers, professional and technical associations; and industry groups.
- *Failure mode and effect analysis (FMEA).* An analytical procedure in which each potential failure mode in every component of a product is analyzed to determine its effect on the reliability of that component and, by itself or in combination with other possible failure modes, on the reliability of the product or system and on the required function of the component; or the examination of a product (at the system and/or lower levels) for all ways that a failure may occur. For each potential failure, an estimate is made of its effect on the total system and of its impact. In addition, a review is undertaken of the action planned to minimize the probability of failure and to minimize its effects.
- *Fast tracking.* A specific project schedule compression technique that changes network logic to overlap phases that would normally be done in sequence, such as the design phase and construction phase, or to perform schedule activities in parallel. See also crashing.
- *Finish-to-finish (FF).* The logical relationship where completion of work of the successor activity cannot finish until the completion of work of the predecessor activity.
- *Finish-to-start (FS).* The logical relationship where initiation of work of the successor activity depends upon the completion of work of the predecessor activity.
- *Firm-fixed-price (FFP) contract.* A type of fixed-price contract where the buyer pays the seller a set amount (as defined by the contract), regardless of the seller's costs.
- *Fixed-price-incentive-fee (FPIF) contract.* A type of contract where the buyer pays the seller a set amount (as defined by the contract), and the seller can earn an additional amount if the seller meets defined performance criteria.
- *Fixed-price or lump-sum contract.* A type of contract involving a fixed total price for a well-defined product. Fixed-price contracts may also

include incentives for meeting or exceeding selected project objectives, such as schedule targets. The simplest form of a fixed-price contract is a purchase order.

- *Float.* Also called slack. See also free float.
- *Flowcharting.* The depiction in a diagram format of the inputs, process actions, and outputs of one or more processes within a system.
- *Free float (FF).* The amount of time that a schedule activity can be delayed without delaying the early start of any immediately following schedule activities.
- *Gantt chart.* See *bar chart.*
- *Imposed date.* A fixed date imposed on a schedule activity or schedule milestone, usually in the form of a "start no earlier than" and "finish no later than" date.
- *Influence diagram.* Graphical representation of situations showing causal influences, time ordering of events, and other relationships among variables and outcomes.
- *Integrated change control.* The process of reviewing all change requests, approving changes and controlling changes to deliverables and organizational process assets.
- *Invitation for bid (IFB).* Generally, this term is equivalent to request for proposal. However, in some application areas, it may have a narrower or more specific meaning.
- *Lag.* A modification of a logical relationship that directs a delay in the successor activity. For example, in a finish-to-start dependency with a 10-day lag, the successor activity cannot start until 10 days after the predecessor activity has finished. See also lead.
- *Late finish date (LF).* In the critical path method, the latest possible point in time that a schedule activity may be completed based upon the schedule network logic, the project completion date, and any constraints assigned to the schedule activities without violating a schedule constraint or delaying the project completion date. The late finish dates are determined during the backward pass calculation of the project schedule network.
- *Late start date (LS).* In the critical path method, the latest possible point in time that a schedule activity may begin based upon the schedule network logic, the project completion date, and any constraints assigned to the schedule activities without violating a schedule constraint or delaying the project completion date. The late start dates are determined during the backward pass calculation of the project schedule network.
- *Latest revised estimate.*
- *Lead.* A modification of a logical relationship that allows an acceleration of the successor activity. For example, in a finish-to-start dependency with a 10-day lead, the successor activity can start 10 days

before the predecessor activity has finished. See also lag. A negative lead is equivalent to a positive lag.

- *Life cycle*. See project life cycle.
- *Materiel*. The aggregate of things used by an organization in any undertaking, such as equipment, apparatus, tools, machinery, gear, material, and supplies.
- *Matrix organization*. Any organizational structure in which the project manager shares responsibility with the functional managers for assigning priorities and for directing the work of persons assigned to the project.
- *Milestone*. A significant point or event in the project. See also schedule milestone.
- *Monte Carlo analysis*. A technique that computes, or iterates, the project cost or project schedule many times using input values selected at random from probability distributions of possible costs or durations, to calculate a distribution of possible total project cost or completion dates.
- *Opportunity*. A condition or situation favorable to the project, a positive set of circumstances, a positive set of events, a risk that will have a positive impact on project objectives, or a possibility for positive changes. Contrast with threat.
- *Organizational breakdown structure (OBS)*. A hierarchically organized depiction of the project organization arranged so as to relate the work packages to the performing organizational units. (Sometimes OBS is written as organization breakdown structure with the same definition.)
- *Parametric estimating*. An estimating *technique* that uses a statistical relationship between historical data and other variables (e.g., square footage in construction, lines of code in software development) to calculate an *estimate* for activity parameters, such as *scope, cost, budget*, and *duration*. This technique can produce higher levels of accuracy depending upon the sophistication and the underlying data built into the model. An example for the cost parameter is multiplying the planned quantity of work to be performed by the historical cost per unit to obtain the estimated cost.
- *Pareto chart*. A histogram, ordered by frequency of occurrence, that shows how many results were generated by each identified cause.
- *Position description*. An explanation of a project team member's roles and responsibilities.
- *Precedence relationship*. The term used in the precedence diagramming method for a logical relationship. In current usage, however, precedence relationship, logical relationship, and dependency are widely used interchangeably, regardless of the diagramming method used.

- *Predecessor activity.* The schedule activity that determines when the logical successor activity can begin or end.
- *Product life cycle.* A collection of generally sequential, nonoverlapping product phases whose name and number are determined by the manufacturing and control needs of the organization. The last product life cycle phase for a product is generally the product's deterioration and death. Generally, a project life cycle is contained within one or more product life cycles.
- *Product scope.* The features and functions that characterize a product, service, or result.
- *Product scope description.* The documented narrative description of the product scope.
- *Program.* A group of related projects managed in a coordinated way to obtain benefits and control not available from managing them individually. Programs may include elements of related work outside of the scope of the discrete projects in the program.
- *Program management.* The centralized coordinated management of a program to achieve the program's strategic objectives and benefits.
- *Program management office (PMO).* The centralized management of a particular program or programs such that corporate benefit is realized by the sharing of resources, methodologies, tools, and techniques, and related high-level project management focus.
- *Project.* A temporary endeavor undertaken to create a unique product, service, or result.
- *Project charter.* A document issued by the project initiator or sponsor that formally authorizes the existence of a project, and provides the project manager with the authority to apply organizational resources to project activities.
- *Project life cycle.* A collection of generally sequential project phases whose name and number are determined by the control needs of the organization or organizations involved in the project. A life cycle can be documented with a methodology.
- *Project organization chart.* A document that graphically depicts the project team members and their interrelationships for a specific project.
- *Project scope statement.* The narrative description of the project scope, including major deliverables, project objectives, project assumptions, project constraints, and a statement of work, that provides a documented basis for making future project decisions and for confirming or developing a common understanding of project scope among the stakeholders. A statement of what needs to be accomplished.
- *Resource leveling.* Any form of schedule network analysis in which scheduling decisions (start and finish dates) are driven by resource

constraints (e.g., limited resource availability or difficult-to-manage changes in resource availability levels).

- *Responsibility matrix.* A structure that relates the project organizational breakdown structure to the work breakdown structure to help ensure that each component of the project's scope of work is assigned to a responsible person.
- *Risk.* An uncertain event or condition that, if it occurs, has a positive or negative effect on a project's objectives.
- *Risk acceptance.* A risk response planning technique that indicates that the project team has decided not to change the project management plan to deal with a risk, or is unable to identify any other suitable response strategy.
- *Risk avoidance.* A risk response planning technique for a threat that creates changes to the project management plan that are meant to either eliminate the risk or to protect the project objectives from its impact. Generally, risk avoidance involves relaxing the time, cost, scope, or quality objectives.
- *Risk breakdown structure (RBS).* A hierarchically organized depiction of the identified project risks arranged by risk category and subcategory that identifies the various areas and causes of potential risks. The risk breakdown structure is often tailored to specific project types.
- *Rolling wave planning.* A form of progressive elaboration planning where the work to be accomplished in the near term is planned in detail at a low level of the work breakdown structure, while the work far in the future is planned at a relatively high level of the work breakdown structure, but the detailed planning of the work to be performed within another one or two periods in the near future is done as work is being completed during the current period.
- *Root cause analysis.* An analytical technique used to determine the basic underlying reason that causes a variance or a defect or a risk. A root cause may underlie more than one variance or defect or risk.
- *Schedule milestone.* A significant event in the project schedule, such as an event restraining future work or marking the completion of a major deliverable. A schedule milestone has zero duration. Sometimes called a milestone activity. See also milestone.
- *Scope.* The sum of the products, services, and results to be provided as a project. See also project scope and product scope.
- *S-curve.* Graphic display of cumulative costs, labor hours, percentage of work, or other quantities, plotted against time. The name derives from the S-like shape of the curve (flatter at the beginning and end, steeper in the middle) produced on a project that starts slowly, accelerates, and then tails off. Also a term for the cumulative likelihood

distribution that is a result of a simulation, a tool of quantitative risk analysis.

- *Statement of work (SOW)*. A narrative description of products, services, or results to be supplied.
- *SWOT analysis (strengths, weaknesses, opportunities, and threats analysis)*. This information-gathering technique examines the project from the perspective of each project's strengths, weaknesses, opportunities, and threats to increase the breadth of the risks considered by risk management.
- *Triple constraint*. A framework for evaluating competing demands. The triple constraint is often depicted as a triangle where one of the sides or one of the corners represents one of the parameters being managed by the project team.
- *Value engineering (VE)*. A creative approach used to optimize project life cycle costs, save time, increase profits, improve quality, expand market share, solve problems, and/or use resources more effectively.
- *Work breakdown structure (WBS)*. A deliverable-oriented hierarchical decomposition of the work, to be executed by the project team to accomplish project objectives and create required deliverables. It organizes and defines the total scope of the project. Each descending level represents an increasingly detailed definition of the project work. The WBS is decomposed into work packages. The deliverable orientation of the hierarchy includes both internal and external deliverables.

Index

For Product Safety Concerns and Information please contact our EU
representative GPSR@taylorandfrancis.com Taylor & Francis Verlag GmbH,
Kaufingerstraße 24, 80331 München, Germany

Printed and bound by CPI Group (UK) Ltd, Croydon, CR0 4YY
08/05/2025
01864414-0005